Selling Information Technology

Life in the IT Industry Spanning Five Decades.
The Story of My Career.

-a memoir-

Chuck Cliburn

Parker House Publishing

Copyright © 2021 New Capitol IT, LLC.

All rights reserved worldwide.

No part of this book may be reproduced in any form or by any electronic or mechanical means, including information storage and retrieval systems, without written permission from the author, except for the use of brief quotations in a book review.

This book is designed to provide information and inspiration to our readers. The content is the sole expression and opinion of the author. Neither the publisher nor the author shall be liable for any physical, psychological, emotional, financial, or commercial damages, including but not limited to special, incidental, consequential, or other damages.

Disclaimer
This book is a memoir of my career, beginning with my college years. The stories are true and accurate to the best of my memory. The people in the book are real people, but some of the names have been changed. The book also contains a good number of my opinions and views on various things. My opinions and views are opinions and views – nothing else. All dialogues in quotes have been recreated to express the nature of the dialog as accurately as possible.

ISBN: 978-1-7372856-0-1

Cataloging in Publication Data is on file at the Library of Congress

Cliburn, Chuck
Selling Information Technology: Life in the IT Industry Spanning Five Decades. The Story of My Career.
Memoir
Summary: *Selling Information Technology* provides the reader with an inside view of life in this complex and competitive industry beginning in 1974 and spanning the next four decades. It speaks to the struggles of once-mighty computer companies to simply survive as computers changed from highly proprietary money-makers to open-system commodities in the late eighties and nineties. Those that have survived have done this by adjusting to a world with an insatiable appetite for information and a decreasing appetite for "computers."

Printed in the United States of America

Dedication

For Madelin

Thank you for always allowing me to follow my dream and for being with me every step of the way – especially Philadelphia.

And thank you too for listening to my work stories during the early years. Here are the rest of them. Enjoy!

Love Always,
Chuck

Foreword

I do not know Chuck Cliburn as a salesman, businessman, or lobbyist. Instead, I know him as Pop Pop. He is an amazing musician, plays in a rock band, and writes songs. He likes to go on walks with his dogs and watching his noisy and crazy Cockatiel bird that he and Nana found in their yard. He loves telling me stories about his past and has told me many stories (some more than once) about growing up in Mississippi. I love my Pop Pop and know that you will love his stories just as dearly I do!

Karis Cliburn

Table of Contents

To Be Near Madelin 1
Hey Nineteen! 11
A Peaceful Easy Feeling 19
Conversación Número Ocho 23
Interviewing with IBM and Burroughs 31
Big Company, Lots of Money, Suit and Tie 35
This Chapter Ain't No Joke 53
Miami Vice 69
The Reinvention of the 1980s 81
Southbound 95
Return of the Phoenix Connection 111
Digital Has It Now! 125
Tunnel of Darkness 133
The Mission 139
The Kingdom 147
On the Hook for Saddlebrook 153
A Bridge Too Far 157
When Your Champions Leave 163
The Unthinkable 167
So, You Want to Be in Sales? 169
Welcome to Tallahassee 173
The Light Returns 177
Big Game Hunting 185
Richard Gaddy 195
The Path Out of Hardware 201
Sometimes You Lose, Sometimes You Win Twice 205
Marrakesh, Mysterious Ways 215
The State Technology Office (STO) 223

Space Invaders	233
Everything Was Big	239
Something Good for Florida	253
"They Went Up"	259
Change, The Only Constant	265
The Beginning of the End of ACS	271
The Final Frontier - Florida Enterprise Email	275
New Capitol IT, My Days as a Lobbyist	285
About the Author	289

1

TO BE NEAR MADELIN

I was sitting with Madelin in her Volkswagen parked outside Neshoba Hall, a men's dorm at East Central Junior College (ECJC) in Decatur, Mississippi. Lots of guys were walking by, staring at Madelin, as college guys do. There was nothing left to do but kiss her goodbye, grab my suitcase, and find my dorm room on this muggy Mississippi afternoon. It was late August of 1969 and so began my freshman year of college. I was 18.

Everybody there just called the school "Decatur." The town, Decatur, was small and very rural, with maybe two or three stoplights at most. It had one small café that everyone called "the steakhouse," but that was not its real name, and it did not serve steaks.

My family had moved from Meridian, Mississippi, to Fort Walton Beach, Florida, in June of 1968 before my high school senior year – a bad time for a kid to have to move. From my teenage point of view, Fort Walton Beach was an upgrade over Meridian, and Choctawhatchee High was an upgrade over Northeast Lauderdale High. But I missed my friends and especially my girlfriend, Madelin Pratt. She would be attending Meridian Junior College (MJC) after high school, so my college plan was to be near Madelin, and nothing else mattered much. That was the extent of my plan.

I was at Decatur only because MJC did not have dorms, and Decatur did. It was only 30 miles or so from Meridian, so it was close enough to hitchhike on the weekends and for Madelin to drive over occasionally to pick me up.

I started college with the idea of becoming some kind of engineer, but it was a vague goal at best. I had never known an engineer and had only a basic idea as to what engineers did. But in the late '60s, most of the high school teachers and counselors encouraged the kids that were even halfway decent in math and science to pursue a math or engineering degree – probably due in part to all the NASA stuff. Since I qualified as halfway decent, that's what I did.

After two semesters of chemistry, calculus, and other courses that were boring and too abstract for my big-picture brain, I changed my mind and decided that business was for me.

However, it didn't take that long to change my mind about Decatur. After only a few weeks, I knew that I had made a mistake. Decatur was not going to work out for a whole litany of reasons.

For starters, hitchhiking to Meridian every weekend was a real drag – much worse than I had envisioned. Additionally, I just did not fit in. Put bluntly, Decatur was a tad too "country" for me, and I grew up in rural Mississippi.

Another problem was the seemingly disproportionate number of student thieves. The main band was a group of guys from Scott County, Mississippi. They were friendly enough and funny as hell – but they were thieves. Stuff was continuously being stolen from dorm rooms, mostly books and radios, but really anything that was not nailed down.

The first thing I had stolen was my Kodak Instamatic camera. They never talked about stealing things, but their dorm rooms were full of things like Lions Club gumball machines, street signs, and things that were obviously stolen. They later stole my chemistry book and slide rule but, being "decent thieves," they waited until the semester was over to do that.

One night, the Scott County boys came in the dorm laughing and being obnoxiously loud and obnoxiously drunk. I stuck my head out the door to see what was going on and one of them yelled down the hall; "Hey, Cliburn, come see the Volkswagen we put on top of a fire hydrant." My goal was to finish the semester and get the hell out.

My best friend from high school and later my best man, Leon McKee, was attending MJC with Madelin along with some of our other high school classmates from Northeast Lauderdale High School. He was a terrific guy and was like a brother to me. He was popular in high school, a student council leader, a class officer, and a talented baseball and football player. Everyone loved him.

He turned 18 in October of 1969 while I was at Decatur. He had sent me a postcard telling me he was coming over on the day of his birthday to celebrate. I think it may have been a Friday because I remember the campus being pretty much empty. We always had fun, usually just talking and laughing, so I was happy that he was coming.

Just before dark on the day of Leon's birthday, there was a knock on my dorm door. It was Leon holding a bag of ice and a 6-pack of beer. While this may seem quite normal for two 18-year-old college kids, it was not normal for us. Neither one of us had ever drunk a beer or anything alcoholic.

We both had plenty of opportunities, but because of our upbringing and respect for our parents who did not drink (and fear to a certain degree), we did not drink. Sooner or later though, it had to happen.

We put the beer in my sink, dumped the bag of ice on it, and went to get something to eat. It was only fitting that I drink my first beer with my best friend since the 8th grade – but it would not be this night. We talked about it and decided not to drink the beer after all. We sold it to some guys in the dorm at a premium price.

It was not over though. Several months later I drank my first beer with my good friend Leon. After that, drinking beer with Leon and other friends became "a thing."

I transferred to MJC after one semester at Decatur – not a day too soon. I rented a room in a boarding house in downtown Meridian for next to nothing. There were plenty to choose from. These were old, creepy, large homes usually owned by an elderly couple or a widow that lived in a portion of the house. I lived in three different ones during my time there. They were always easy to find with permanent signs in the yards: "Rooms - $8.00 per week." It worked out fine.

For the most part, I was always with Leon or Madelin. I didn't have a car, so Madelin or Leon would pick me up every morning at my boarding house for the ride to MJC. At the end of the day, someone would give me a ride home.

Leon was no different at MJC than he was in high school – he quickly became popular. Even though he was from a county school, not Meridian High like most of the in-crowd, he quickly charmed everyone from the in-crowd students to the MJC faculty and staff. He played baseball for MJC and was a student leader.

We were much alike in many ways, but quite different in others. I was quite shy and not much of a socializer. Leon, on the other hand, never saw a group that he did not want to belong to. He had a different date almost every weekend too, while Madelin was the only date I wanted.

I recall being in the cafeteria one day with Leon shortly after the semester started. We had been through the line with our trays and were looking for a table. I followed Leon to a table where a man dressed in a suit was sitting. "Mr. Bahr, do you mind if we join you?" Leon asked. The man said, "Hi, Leon, please do." We sat down and Leon introduced me to Richard Bahr, the Dean of Student Services. It was obvious that they knew each other. Mr. Bahr was nice to me and asked me lots of questions including how I came to be at MJC, what I wanted to do with my career, and so forth. From then on, he always treated me as a friend and always used my name when he spoke to me when we would occasionally see each other in the hallway.

Leon told me that he also wanted to introduce me to "Turner." Turner was Mr. John Turner, the head academic counselor. We went to his office and Leon introduced me in glowing terms – much better than I deserved and much better than was accurate. Mr. Turner was very nice and made me feel welcome. When we left, Leon told me that Turner had a cool apartment and would occasionally invite a few students over to hang out. While Leon didn't say so, the "few students" were a small group of the in-crowd, and Leon was one of them.

Sure enough, Madelin and I were soon invited to one of the informal hangout parties at Turner's. We went and had a great time hanging out with the MJC in-crowd. It was a cool apartment for sure, but I was mainly impressed with his console stereo system playing one of the new Chicago albums.

After graduating from high school, I had saved enough money from my summer job as a dishwasher to pay for my first-semester tuition, dorm, and meals at Decatur. Unfortunately, I was broke by the beginning of the second semester so my dad paid my tuition for my first semester at MJC. It was up to me to pay for my food, my room at the boarding house, and my miscellaneous expenses. Since I didn't have a car, I was not sure how I was going to do this, but I had learned from my mom and dad to have faith, that God would provide a way – and He did.

Leon told me that MJC had a student work program where students could work part-time for the school or the Meridian Public School District. One of the work programs was with the Meridian Public School District where MJC students worked as recreational assistants at elementary schools. The informal name of the program was "play teaching." Leon was already in the program and was a "play teacher" at Poplar Springs Elementary in the afternoons. He told me I should apply and try to get assigned to the same school. That way, I could ride with him every day.

He introduced me to the director of the program. I applied, got the job, and was assigned to the same elementary school. It was perfect.

For the rest of the semester, I was a "play teacher" working with Leon at Poplar Springs Elementary. We were basically recess coaches. To the kids, I was Coach Cliburn. It was great fun, but I learned right away that I did not want to be an elementary school teacher.

I also had a second job through the school as a language lab operator at night. That was great too. All I had to do was to sit at the console, put in language tapes for the students, and log in their time. Most of the time, the students, especially the girls, brought in a favorite rock and roll cassette and asked me to play it instead of the required language tape. Abbey Road was the most popular and I was always happy to help. I was there for two hours each night and used the time to study while listening to my favorite rock music.

I went home to Fort Walton Beach for the summer to work so I could save enough money to at least pay my tuition for the upcoming fall semester. I went back to the same hotel, got hired by the same chef, but got a better job this time – cook assistant. The hotel had new owners and had changed its name to the International Beach House.

The chef that hired me as a dishwasher the summer before, Eddie Jesse, was still there. Eddie was a kind old man that reminded me of grandpa on The Munsters. I worked there for three summers and he never once called me "Chuck." To him, I was "Sonny Boy." "Sonny Boy, come over here so I can show you how to bread the shrimp," he would say. He was a practical joker too. He once walked over to me with a heaping spoon of something and said, "Sonny Boy, taste this sauce for me." He dumped the whole spoonful in my mouth. It was horseradish – he thought that was funny as hell, and so did the rest of us.

During my first summer there, I couldn't wait for a busboy or a waiter to leave so that I could hopefully move "up" to one of those jobs. Those guys wore red jackets, white shirts, and black bow ties. Plus, they didn't work half as hard I did. I wore my most-busted jeans, tee shirts, and kitchen aprons.

Within a few weeks, one of the busboys quit and the restaurant manager came into the kitchen to ask me if I would like to have that job. I was thinking, "Hell, yeah" but probably said, "Yes, sir, I sure would." The busboys and waiters didn't work for Eddie, so the manager told me he would talk with Eddie to get his ok and get back to me.

The next night, he came back into the kitchen and told me that Eddie would have nothing to do with it. Eddie had told him that "Sonny Boy" was the best dishwasher that he had ever had and that they could go find their own busboy. I was disappointed for sure but happy to know that I was such a good dishwasher. Maybe that's why Eddie hired me as a cook assistant the following summer and gave me a 25-cent-per-hour raise. It certainly was not because I knew how to cook.

I worked in the kitchen for three summers and learned to cook pretty much everything on the menu. At times, I was the head breakfast cook – something that Madelin still has a hard time envisioning. Looking back, so do I.

I returned to MJC in the fall of 1970 in my first car, a '65 GTO. I knew it was cool, but I did not realize that it would end up being the coolest car I would ever own. My dad helped me get a bank loan with a payment of $45 per month. With my own car, I now had the flexibility to take off-campus jobs.

The school's placement center told me of an opening for a part-time house trailer salesman. Today they are mobile homes but back then they were house trailers, at least in

Mississippi. I interviewed and got the job.

The same man that owned the house trailer company, Trailer Sales, Inc., Mr. D.A. Toney, also owned a pulpwood company in Meridian, Shubuta Tie and Timber. Mr. Toney's General Manager was a man named Dees. They were always together.

After being on the job for a few months, Dees drove up and came into the office just as I was closing up. He told me that he was sorry, but they were hiring a second full-time sales guy and I would no longer be needed – but he was nice about it. I felt that was kind of weird because I knew that Mr. Toney liked me. I was hurt by it too. It felt like being dumped by a girlfriend.

The next afternoon I was in the MJC library studying and sensed someone standing over me. I looked up and it was Mr. Toney. The best way for him to reach me was to look for me at the college and that is what he did. He quipped that, if I studied too much, I would get a headache. He then quietly asked me if wanted my job back. We walked outside.

He told me that Dees was not supposed to let me go and that he was not sure why that happened. However, I have always felt that Dees probably did what he was told to do and that Mr. Toney just felt guilty about it.

In any case, he explained that he needed a "utility guy" doing work for both of his companies and that it was my job if I wanted it. Sometimes, I would fill in at the trailer dealership. Other times, I would assist his bookkeeper at the timber company, and sometimes I would drive around southern Mississippi delivering various papers and things to his other house trailer dealerships and timber yards.

I immediately gave him my best "Yes, sir, and thank you."

He then told me that if I wanted to start "right now," I could drive to McGee to deliver some papers that afternoon.

Within 15 minutes, I was on my way to McGee, Mississippi, computing in my head how much I was going to make at $1.65 per hour plus 10 cents a mile. I worked for Mr. Toney for the rest of my time at MJC and have always been grateful for the help he gave me.

Madelin, Leon, and I graduated from MJC in 1971, but they had to go without me to the graduation ceremony. I never attended any of my graduation ceremonies, except high school, and I participated in that one only because Madelin came from Meridian on a Greyhound bus to see me graduate.

Madelin and I were married on September 4, 1971, and moved to Hattiesburg where I attended the University of Southern Mississippi (USM or Southern). Leon transferred to Mississippi State where he graduated with a degree in forestry. We would see him occasionally on our visits back to Meridian and at occasional high school reunions, but our lives gradually drifted apart.

Leon had a successful career in the timber industry, never leaving Lauderdale County, Mississippi, that I know of. I guess you could say his dream career came to him while I moved all over the country chasing mine.

Sadly, Leon died much too early from brain cancer when he was only 55. He was laid to rest at the Andrew Chapel Methodist Church cemetery near Meridian – the church where my dad was a pastor and where I first met Leon and Madelin when we were 12. I have had some very good friends since, but never anything quite like the special friendship I had with Leon during my teen years.

2

HEY NINETEEN!

Being nineteen was a big year for Madelin. She became the only 19-year-old Registered Nurse I have ever known, and she married me! Shortly after we arrived in Hattiesburg, she was hired as a pediatric RN at Forrest General Hospital and went on to become the best nurse anyone could ever imagine.

I enrolled as a junior at Southern. Not knowing exactly what I wanted to do, I chose Personnel Management as my major. It sounded cool to me, but that was about it. Everyone calls it Human Resources now, which I guess somebody decided sounds better.

During my first semester, I took an introductory marketing class taught by Dr. Richard Vreeland, the head of the Marketing Department. He was a fascinating man who loved to talk about everything except marketing. He loved paranormal stuff, parapsychology, and séance kind of stuff. He talked often about Edgar Cayce, a self-professed clairvoyant of the early 20th century.

Edgar Cayce's theories and beliefs were a common theme to many of his lectures. He talked a lot about the Age of Aquarius too. He took a lot of pleasure in explaining to us that it was more than just a song by The 5th Dimension. I was in my third year of college and this was the first class that met my expectation as to what college should be. It was a million times more interesting than the one personnel management class I took, but probably because Dr. Vreeland never talked much about marketing.

But I especially liked and admired him. So much so that I soon asked him if I could meet with him to discuss possibly changing my major. He was gracious and told me to make an appointment with his office. I met with him soon afterward.

It was this meeting that set the path for my career.

He was very helpful, just as I thought he would be. I shared with him that I had chosen personnel management for no real reason, maybe because it had "management" in the title. As you would expect, he asked me what I wanted to do for my career. I told him that my main goals were to get a good job with a big company, to make a lot of money, and to wear a suit and tie to work.

Not surprisingly, he suggested that I consider changing my major to marketing. He explained that there were several career paths in marketing, including advertising and consumer research, but that most of the good jobs were in sales – and so was most of the money.

He showed me a computer printout on wide green-bar computer paper that was a report of jobs placed through the USM placement center. Two things were clear. He knew how to convince students to major in marketing, and there were, indeed, a lot of good jobs going to marketing majors.

Most of these jobs were in sales, predominantly with heavy equipment, pharmaceutical, and computer companies. He shared with me that a previous student of his was making $50,000 a year after three years as a salesman with a steel manufacturer and that another marketing major was making $20,000 a year after two years with a fabricated home manufacturer. I grew up as the son of a country preacher and had never known anyone that made that kind of money. The average starting salary out of college back then was around $10,000 a year or so. I was convinced and changed my major to marketing.

For the next two years, I took a lot of marketing courses and enjoyed them all. I signed up for every class I could that was taught by Dr. Vreeland.

Another professor that I really liked was Dr. Tom Ivey. Dr. Ivey was a psychologist who taught in the business school. He taught marketing classes like consumer behavior and consumer research. He was smart and entertaining. He looked like a shrink, talked like a shrink, and was, well, a shrink. After one class with him, it made perfect sense as to why a psychologist would teach consumer behavior classes. After all, why people buy one brand of shampoo over another has more to do with psychology than anything else.

Dr. Ivey, a graduate of Arizona State University, spoke often and in glowing terms about the Phoenix area, Arizona State, and a private graduate school in Glendale, Arizona, the American Graduate School of International Management, known by most as "Thunderbird."

Dr. Ivey did not attend Thunderbird, but he was a big fan. He mentioned Thunderbird often, especially in any discussion about international business. He believed that Thunderbird was the best international business graduate school in the country and would often cite independent college rankings and various articles to support his view.

I took quite a few of his classes. Due mainly to his influence, I would eventually attend and graduate from Thunderbird.

Another USM marketing professor of note was Dr. Richard Stevens. He was somehow related to Dr. Ivey – I think his brother-in-law. He was a good professor too, but unlike Dr. Vreeland and Dr. Ivey, he was as boring as the day was long. Anyway, I took one marketing course from him and it was one of my more memorable college experiences, albeit not a good one.

I usually took at least one night class per semester and this was one of them. I found the night classes to be more relaxed, smaller in size, and usually easier. One of these was a marketing course taught by Dr. Stevens. There were three guys in the class that I knew, and the rest were mainly older students that probably had full-time day jobs. The four of us studied together and sat together.

When it came time for finals, one of the guys excitedly told the rest of us that someone had given him a folder with a "shit load" of old tests. We agreed to meet at his apartment to review these old tests and to use them to prepare for the final exam.

He made copies for us and handed them out when we got to his apartment. One of the tests was a hundred-question multiple-choice test – but with no answers. The professor had told us that the exam would be "multiple-choice" or, as we called it, "multiple-guess." We decided that the best course of action would be to use the 100-question test as our study guide reasoning that surely many of these questions would be on the exam. We divided up the questions and spent the next couple of hours looking up the answers. Once that was done, we all memorized the correct answers.

When we got to class for the exam, the professor handed out the exams (always paper tests in those days) and to the shock and joy of us all, it was the exact same exam. We glanced up at each other trying to conceal our exuberance. After completing the exam and waiting for someone other than us to finish, we left the classroom and waited for each other in the building lobby for the self-congratulatory high fives.

A couple of days later, one of the guys called me late one afternoon with panic in his voice to ask me if I had been by Dr. Stevens' office. The professors always posted the test scores on the hallway wall outside their offices, so this is what I assumed he was talking about. He told me to go look and meet them in the library. I jumped in the car and headed to the business building only ten minutes away.

I walked down the long hallway to Dr. Stevens' office with a little dread and lots of curiosity, wondering what this could be about.

Dr. Stevens had taped all four of our exams to the wall with a red circle around the only answer that we all got wrong. Out of 100 questions, we each missed only one question. We all missed the same question with the same wrong answer. Dr. Stevens did not comment but taped our tests on the wall for us and the rest of the world to see.

I walked quickly to the library to meet the other guys. We went outside so that we could talk and immediately agreed that we were in deep shit. The only issue was "what to do."

We settled on calling Dr. Ivey to get his opinion. He was friendly to all of us and was Dr. Stevens' brother-in-law. We called him and explained exactly what had happened. He already knew about the whole thing and seemed to find it quite amusing. Anyway, he said we should go see Dr. Stevens the next day and explain to him exactly what happened. So that's what we did.

Scared as hell, we all went in together. After explaining everything, he said that he knew something was up, he just did not know what. He told us that it was "on him" for not being more careful with his exams but that it would not be right for us to be the only students in the class to get A's under this circumstance. He let us know that we were the only students to score higher than 90 on the exam and thanked us for coming in and for being honest.

We left the meeting not knowing what the outcome would be. Would he give us F's, or maybe incompletes? We just did not know. When the final grades came out, we all got C's. While most of us would have probably made at least a B had this not happened, we accepted our C's and moved on.

As I got closer to graduation, I began to consider staying at USM to get an MBA, but the idea of going to Thunderbird would not leave me. At first, it seemed more like a dream than a realistic option. Staying in Hattiesburg was one thing, but moving to Arizona with Madelin and our dog, Dube, was something else again.

Madelin and I discussed it a lot and she was completely onboard. I talked about it with my dad, Dr. Vreeland, and Dr. Ivey as well. Everyone was all in, so I ordered a Thunderbird catalog and all the material they could send me. I applied and was accepted to start in the fall term of 1973.

Selling Information Technology

I graduated from Southern in February of '73 so I needed a job from March until August and was content with the probability of working in a restaurant or grocery store, or something like that until we moved to Arizona.

However, I had the opportunity to interview with General Electric Credit Corporation for a management trainee position based in Hattiesburg. I rationalized that I may like it and may even want to stay and continue a career with GE Credit instead of going to grad school. But I was only rationalizing about taking a job that I knew I would not keep. I interviewed and got the job. It was a real job and a great experience that looked good on my resume later. I even had a company car.

In August I resigned, and we moved to Phoenix. I did feel a little guilty, but I got over it soon.

Chuck Cliburn

3

A PEACEFUL EASY FEELING

Other than having our two children, moving to Phoenix was the single most exciting thing that ever happened for Madelin and me. It was the beginning of an adventure that would last a lifetime.

August of 1973 was one of the most fun and most memorable months of our lives. Early that month, I took our Ford Gran Torino to Sears in Hattiesburg to have a trailer hitch installed – even that was fun. A few days later, we rented a small U-Haul trailer that easily held everything we owned with room to spare. That was fun too.

We loaded up and headed to Phoenix with our first stop in Meridian to spend the night with Madelin's parents and to say goodbye.

We got up early the next morning to get an early start. It was a beautiful summer day. There were lots of hugs with Madelin's dad and mom, and plenty of tears too. I don't think either of us realized at the time how hard it was for our parents to see us move so far away. I believe everyone, including us, instinctively knew that we would never live in Mississippi again.

Finally, we were on our way, headed west on I-20. It took us three days to get to Phoenix.

We spent the first night in Dallas and the second night in El Paso. We stopped often to take pictures, especially after we were west of Dallas. We listened to the radio all the way and it seemed that the Allman Brothers' Ramblin' Man was playing constantly. It became our anthem. To this day, when we hear Ramblin' Man it puts us in our Gran Torino driving to Phoenix.

We reached the Arizona state line around mid-morning of the third day and stopped to take a lot of pictures of the Welcome to Arizona sign. Just as I had envisioned, Arizona was the most beautiful place I had ever seen. Admittedly, I had not seen much. Madelin loved it too but maybe not quite as much as I did. We arrived in Phoenix late that afternoon.

We had rented an apartment, sight unseen, from an older couple that were friends of Madelin's parents. They lived in Meridian but were from Phoenix and still owned a house there with an attached apartment that was perfect for us. It was in an older residential neighborhood near downtown and only two blocks from an outpatient surgery facility, SurgiCenter, where Madelin would eventually work, and two blocks from Good Samaritan Hospital where our son, Brian, would be born in 1977.

We unhitched the trailer, unloaded a few things, and met our neighbors, the Shavers. They were expecting us and had the keys to our apartment. They were in their early '80s, very nice, and looked after us like grandparents.

We still had a little time before dark so we drove out to

the Thunderbird campus in Glendale where I would spend most of my time for the next year. The campus was on the property of a World War II airfield and many of the original buildings were still in use. We walked around some, but it was hot as blue blazes. I was excited to finally be there, but a little intimidated at the same time.

We stopped at a convenience store on our way back to call our parents from a pay phone to let them know that we were safe and sound in Phoenix before heading to our apartment for our first night in Arizona. Even though we would be in Arizona for only four years, Arizona would be in us for the rest of our lives.

We absolutely loved it. To us, Arizona was more than just a place. It was a feeling and the door to a life that we never imagined before moving there. To borrow from the Eagles, our life in Phoenix was a peaceful easy feeling. We were young adventuresome free spirits living in the middle of one of God's wonders.

When I hear the Eagles sing "I want to sleep with you in the desert tonight, with a billion stars all around" it puts me in the Arizona desert, camping with Madelin as we often did. I started my career there, we bought our first home there, and our first child, Brian, was born there. Arizona would be forever etched in our souls.

Chuck Cliburn

4

CONVERSACIÓN NÚMERO OCHO

Thunderbird is a global business management school that was founded in 1946. In 2014, it became part of Arizona State University, but for most of its existence, it was a private school dedicated to graduate-level degrees in international business. When I was there, it had recently changed its name from the Thunderbird Graduate School of International Management to the American Graduate School of International Management – but everybody still called it "Thunderbird." The school eventually put Thunderbird back in its name and it is now the Thunderbird School of Global Management.

Thunderbird has always been well-known in international circles for its focus and dedication to international business. However, had it not been for Dr. Ivey, I probably would have never heard of it. But I was there, and getting my Master's degree was now within reach.

Getting this degree was one of the most difficult things I have ever done and the only time in my life when I considered giving up. It was the language requirements that were so difficult for me. With Thunderbird being an international business school, language was an integral and major part of the curriculum.

The requirement was to be fluent, or near fluent, in at least two languages upon graduation. However, if you were already fluent in English and at least one other language, you were not required to take any language courses at all, and many of the students were bilingual. The closer you were to being bilingual, the fewer language courses you were required to complete.

We were immediately given a language assessment test to determine our competency level and where we would begin in the language curriculum. I chose Spanish, thinking it would be easier than most.

My assessment took only a few minutes. After about three questions, I was told that I would start in beginning Spanish. No shock there. I had never taken a Spanish class, not even in high school.

Conversational Spanish met every day and Spanish grammar met twice a week on Tuesdays and Thursdays. Both were extremely intensive.

The conversational classes were small, with no more than eight students. On the first day of class, our instructor told us that we would not bring pencils and paper to class – there would be no reading and no writing. Everything would be spoken. He then paired us for the first week and told us to learn "conversation one" from our cassette tapes to recite for him in class the next day. These were dialogue conversations with no English translation and no corresponding written text.

Each day we would memorize a new conversation, both

sides of the dialogue, and each week we would change partners. This is all we did for the first semester. Of course, I had a full load of business classes too – international business law, global economics, finance, accounting, marketing, and stuff like that.

After class, I went to the bookstore to buy the cassette player, the tapes, and all the books needed for my other classes. I was not concerned at all and even thought that not taking paper and pencils to class was a sure sign that this would be an easy class. Never had I been so wrong.

I got home late that afternoon and told Madelin all about my first day, and there was much to tell.

I decided to memorize my dialogue first, just to get that out of the way before starting my other assignments. I pulled "tape one" from the box of cassette tapes, put it in the player, and pushed play. I was terrified. The first three words were "conversación número ocho." I assumed that "conversación" meant conversation, that número meant number, and that ocho meant one. I was off to a horrific start. Those were the only three words that I could identify as words.

I listened to the entire conversation which lasted for only two or three minutes. Honestly, it sounded like one giant Spanish-garble word. I literally could not tell where one word stopped and where the next word began. To make horrible worse, it was spoken at the normal breakneck Spanish-speaking pace. I still do not understand how anyone can speak that fast, but maybe that's just the southern in me.

For the next eight hours, I memorized this three-minute Spanish-garble word – one syllable at a time. I will never forget the first sentence. *Son extranjeros esos alumos?* I had absolutely no idea what any of this meant. Also, it is not what

I heard and it's certainly not what I said. I thought the first word of the sentence was sonic, as in sonic boom, and that the next two words were words sounding like "train arrows" – well, you get the idea. *Sonic train arrows eso salumunos* or something close to that.

I got up early the next morning and drove to the campus feeling very worried about this dialogue. Conversational Spanish was my first class of the day. The professor walked in, said, "*Buenos Dias*," pulled his chair up right in from of my partner and me, and pointed at my partner to go first. My partner seemed just as nervous as I was and not much better at Spanish either. He began.

Whatever he said did not sound anything close to "train arrows." Undaunted, I did my best to recite the second line of the Spanish garble dialogue. The instructor chuckled. He said, "Senor, when we get to conversation number eight, it will be easy for you." I had learned conversation number eight. It turns out that *ocho* is "eight," not "one" – who knew?

So, our dialogue started something like this:

"I like your car, what is it?"

"No, those students are from Arizona."

But at least it relieved the stress for me for the moment and it helped my partner get off the hook too for the first day. The instructor went on to the other students for them to recite their dialogues. Tomorrow would be another day.

After class, I asked several of my classmates how long it had taken them to learn the dialogue. You know, misery loves company. To my dismay, most of them said 30 minutes to an hour or so. One of them told me he learned it in his car on the drive to school. I was so disheartened and discouraged. How could this be?

There were only eight of us in the class and we came to know each other quite well. It turned out that only two of us were completely void of any Spanish knowledge. One student had lived in Spain for much of his childhood. One was a previous Spanish teacher in high school. One had a wife from Mexico who apparently spoke lots of Spanish at home. The others, except one, had similar stories. Even the one guy who was essentially in the same boat with me had taken Spanish in high school.

I carried on, accepting the fact it would take me eight hours, every night, to learn these dialogues.

About three or four weeks in, I told Madelin that I was not sure I could do it and that maybe I should drop out and get an MBA from Arizona State instead. As she has always been, she was very supportive and said that maybe I should keep trying for a while longer.

I met with my professor to talk about it. He was encouraging and told me that I was doing fine. He assured me that it would soon become easier and that he knew how hard I was trying. "Just hang in there," he said.

He then told me something that I did not know and that was a tremendous relief. He said that Thunderbird was aware that the language curriculum put an extraordinary burden on students like me that had little or no proficiency in a second language. He was right about that for sure.

He went on to explain that the grades for all three semesters of Conversational Spanish would be changed, retroactively, to the grade received for the following semesters – assuming the grades were higher. So, if you made a C in the first semester and a B in the second semester, the grade for the first semester would be changed to a B. Likewise, if you made an A in the third semester, the grades for all three semesters would be recorded as A's. What a relief that was!

As the weeks went on, the dialogues gradually became easier to learn and took less time. By the end of the first semester, I was spending probably an hour or two each night learning the dialogues.

The dialogues became more complex in the second semester, but the tapes contained the English translations which made it immensely easier. We also began to read and write papers in Spanish. Toward the end of the semester, we had to make a speech in Spanish on why we came to Thunderbird and what we planned to do with our careers. The third semester was more of the same but with more of a focus on reading Spanish newspapers and speaking on business-related topics.

I made a C for the first semester but my B in the second semester changed it to a B, and my B+ for the third semester resulted in a B+ for all three semesters. "Pluses" added quality points with their grading system. I successfully completed the language curriculum, but it was the most difficult thing I have ever done.

My year at Thunderbird was an awesome experience. I was surrounded by bright and interesting students from all over the world. The professors were incredible, and some were retired executives from Fortune 500 companies. I learned a lot there, but more than anything, I learned the value of not giving up no matter how daunting the challenge may be.

On August 16, 1974, I graduated with a Master of International Management. Years later, I received a letter from Thunderbird informing me that the curriculum I completed had been retroactively approved and accredited for an MBA, so I soon received a new diploma for a Master of Business Administration in International Management. I proudly display both diplomas in my office. From a career and education perspective, graduating from Thunderbird is the one single accomplishment in my life that I am most proud of.

Chuck Cliburn

5

INTERVIEWING WITH IBM AND BURROUGHS

I began interviewing during my third semester with firms that came to our campus and quite a few firms off campus as well.

Madelin and I loved Phoenix and had decided that neither of us was quite ready to move overseas or to an industrial city in the northeast – although we would do exactly that later on. I rationalized that if I joined a large international firm, I could transition to an international position later if I wanted to.

I was particularly interested in the computer industry. It sounded prestigious and would surely be a growth industry for decades to come. Two of the firms that I interviewed with were IBM and Burroughs. Both were large computer companies, and both were recruiting business majors and MBAs for careers in sales.

I interviewed with Burroughs first. The interviews were held at the Burroughs office in downtown Phoenix and went well. All of the salespeople were young college graduates and they all seemed to really like working there. I loved it and felt like I had a good chance to get a job offer.

I interviewed with one of the sales managers, which they called a Zone Manager, with the Branch Manager, who was in charge of the Phoenix office, and with one of the sales guys who had graduated from Thunderbird a year or two ahead of me. His name was Pete Griffen.

Pete and I talked a lot, mainly about Thunderbird, but a lot about Burroughs too. At the end of the interview, he told me that they worked hard but played hard too and that he hoped I would come on board. The other interviews went well too. After being there for half a day, they asked me to fill out an application and told me they would be in touch soon. I was excited about it.

A few days later I interviewed with IBM. The IBM office was also in downtown Phoenix, near the Burroughs building. It was a much different experience than what I had at Burroughs. Everything was formal and very serious, like someone-died serious. Everyone was older too – not old, but not in their 20s. It didn't strike me as a fun place to work, admittedly comparing it to what I had seen at Burroughs. Everyone at the Burroughs office seemed to be having fun, but not so at IBM.

After waiting in the IBM reception area for what seemed like forever, a nice lady came out and escorted me to a conference room. She was holding a folder and a pencil and told me that I would need to take a career aptitude test and then Mr. Somebody would meet with me. She gave me the pencil and the test and told me how much time I had to complete it. I completed the test and took it to her. She thanked me and said that he would be in shortly.

Mr. Somebody soon came in and told me they required all sales candidates to take the technical aptitude test because the sales jobs at IBM were technical in nature. He was not particularly friendly. He got right to the point and told me that I did ok on the test, but that I may want to consider a career in a different industry. I thought he was full of it, and it only made me more determined to work in the computer industry.

Mr. Somebody then said that he wanted to introduce me to someone, which seemed a bit strange since he was clearly not interested in me. He took me down the hall into "someone's" office and it went something like this: "Chuck, this is Jim Somebody Else. Jim, check out Chuck's shoes. Cool, aren't they?" They chuckled a bit and thanked me for coming in.

I had bought a new pair of shoes for my interviews – two-toned brown dress shoes. And yes, they were cool with a mid-'70s flair. I found out later that IBM was all about dark suits, white shirts, and black wingtip shoes.

Still, there is no way to describe Mr. Somebody other than a Class-A asshole. Fair or not, this encounter formed an indelible opinion of IBM with me that has remained until this day. However, I have had some very good friends over the years that worked for IBM. Don't worry IBM, you are a great company.

Within a few days, I received an offer from Burroughs as a Sales and Marketing Trainee in Phoenix, and I accepted it. IBM was now my competitor, and I would compete with them for the next 38 years. I've had a few disappointing losses to IBM for sure. But more often than not, I beat them like a drum, and it was always with the greatest of pleasure.

Chuck Cliburn

6

BIG COMPANY, LOTS OF MONEY, SUIT AND TIE

First, a little about Burroughs. Burroughs was one of the largest computer companies in the world. It had been around for a long time, originally as a manufacturer of mechanical business machines. In the early 1950s, it became one of the first companies to begin work in digital computing technology. Over the next two decades, it became one of the most successful pioneers in the computer industry, especially for the banking and financial industries.

Despite its success and size, it was still dwarfed by IBM. IBM's main competitors at that time were sometimes referred to as the BUNCH – Burroughs, Univac, NCR, Control Data, and Honeywell. There were others too, but mainly it was IBM and the BUNCH. Burroughs merged with Sperry Univac in 1986 and the combined companies changed their name to Unisys.

Burroughs was an established and elite company that manufactured mainframe computers, mid-range computers, and electronic accounting systems for small and mid-sized businesses.

The Burroughs L Series was often referred to as a minicomputer, but it was more of an electro-mechanical ledger card accounting machine with some digital functionality. The Burroughs L Series was the industry's leading accounting machine for a long time but became obsolete a few years after I joined the company.

The company's flagship technology was in its mainframes and its specialized computing equipment designed for the banking industry. The mainframe operating system, MCP, was considered by many industry experts to be the best of its time.

My first day with Burroughs was Monday, August 19, 1974. The office opened at 8 a.m. so I was there at 7:30, not taking any chances of being late. I sat in my car in the parking lot until someone arrived to open the door. When I walked through the door, I had accomplished two of the three goals I had shared with Dr. Vreeland at USM. I was working for a big company and wearing a suit and tie to work. My third goal, making a lot of money, would have to wait.

Everyone was extremely nice and went out of their way to make me feel welcome. The office held a sales meeting every Monday at 8 a.m., so sitting in the sales meeting was my first order of the day and the first business meeting of my career.

One of the guys escorted me to a large conference room which also doubled as one of the demo rooms. Just about everybody walked in with a coffee mug, and someone brought a box of doughnuts. I could tell right away that I was going to like working there. Everyone was talking about their weekend and just seemed relaxed and happy.

When the meeting started, the Branch Manager introduced me and asked if there was anything I wanted to say. I think I said something predictable like how happy I was to be there and that I looked forward to being a part of the team.

Then the meeting started which was basically each sales rep standing up with a magic marker in front of a flip chart listing the deals he was working on, next steps, and when they would close – usually with a few questions from the Zone Manager and Branch Manager. It was a ritual that would, in some form or fashion, be part of my life for the rest of my career.

After the sales meeting, I was given a tour of the building including the other demo rooms and the mainframe computer room. The computer room was a secured area that housed a Burroughs mainframe computer. It had a raised floor, was very cold, and had a loud continuous rumble. They explained that they used the mainframe for various things at night, but that it was leased to one of the Phoenix banks during the day – or maybe vice versa. It was impressive, very big, and had lots of tape drives, giant line printers, and so forth. I was already starting to feel like a big shot.

There were two Burroughs Branches housed in the building – the GA Branch (general accounts) and the NA Branch (named accounts). I worked in the GA Branch. The NA Branch was responsible for specific "named" large accounts, banks, savings and loans, and credit unions, and all mainframe sales. The GA Branch was responsible for everything else. Our product line included calculators, the Burroughs L Series accounting machines, and the Burroughs B700 and B1700 mid-range computer systems.

The rest of the day was consumed with filling out paperwork, reading policy manuals, and meeting people. Before I knew it, I had completed my first day of 23 years with the company. I couldn't wait to get home to tell Madelin about it.

Within a week or two, my boss, Jim Daugherty, was promoted and left Phoenix. That created quite a buzz in the office. The same day that this was announced, the Branch Manager, Bill Savage, called me into his office. Other than my interview, this was the first one-on-one meeting I had with Bill, so I had no idea what to expect. When I walked into his office, Bob Heisser was there with him. Bob was one of the sales managers and was three or four years older than me. They wanted to talk with me about Jim's departure and how it would affect me.

It was all good. Bill told me about Jim's promotion and that they would soon be naming a new Zone Manager to replace him. They were also creating new dedicated sales teams to sell the "group 3S" product line which consisted of the B700 and B1700 computer systems. The new group 3S sales teams would be managed by Selected Account Managers (SAMS).

Bob Heisser was being promoted to the SAM position for the Phoenix Branch. Bob told me that he wanted me to be part of his new team. When I completed my sales training, I would be selling B700 and B1700 computer systems and not the L Series accounting machines like most of the other guys. Also, I would have a much larger territory than the L Series guys and would be on the "ground floor" of the new "group 3S" program. It felt somewhat like a promotion to be selected to sell this larger and more expensive line of computers and to a certain extent, it was.

Bob Heisser was now my new boss.

For the next six months or so, I was in training. I made sales calls with other sales reps, attended formal classes at Burroughs training centers (mostly in Pasadena, California), read a lot, and sold calculators.

Selling Information Technology

The trainees sold calculators to learn the basic concepts of selling. It was a great experience, but not necessarily much fun.

We sold calculators to businesses by making door-to-door cold calls. Well, some of us did. As a trainee, I didn't have a formal territory yet so I would leave the office in the morning with brochures in my briefcase and sometimes with a calculator under my arm to go make calls anywhere I wanted.

One day, I decided to go downtown to try my luck in the high-rise office buildings. I recall walking into one of these buildings and noticing the "no soliciting" sign on the glass door. I ignored it figuring the worst that could happen would be that I got thrown out. If that were to happen, it would be a badge of honor back at the office.

Most of the time, I did not get in to see anyone, but I always left a brochure and my business card. But this day would be different.

I walked into a Johnson and Higgins office with fancy double doors, not even knowing what they did, and introduced myself to the receptionist. "Hi, I'm Chuck Cliburn with Burroughs Corporation. I'm wondering if I could meet with your business manager for a few minutes."

The receptionist did not ask if I had an appointment, as they usually did, and instead said she would go check. She came back in a minute or so and to my surprise said that someone would be out in few minutes. Soon, a man came out, introduced himself, and took me back to his office. I remember that his first name was Larry.

Larry told me what they did and that they had just received approval to buy new calculators for the entire office. I felt like I had won the lottery. He asked me If I could leave some brochures and the calculator for him to use for a couple of days which I enthusiastically did. The only thing I could think to ask was how many they would be buying, to which he said, "at least 20." I was so excited and realized for the first time how much I was going to love selling.

I made an appointment to come back in a few days to get the demo calculator and to hopefully pick up an order. It went even better than planned. I soon ended up with an order for 20 Burroughs calculators and would sell even more to them over the coming months. Getting that order signed was pure exhilaration and it made me somewhat of a hero in the office for selling so many calculators to a new customer.

I soon learned that hardly anyone, at least in our branch, ever really sold calculators. Most of the time they were used to help close the deal on a computer. We would often offer "free software" to help close a computer sale. We did this by offering software at no charge up to the amount the client would spend on calculators as part of the computer deal. At the time, Burroughs was not interested in making a profit on software. It was more of a necessary evil to sell computers. To be fair to the NA guys, they legitimately sold a lot of calculators to the banks and credit unions.

I completed my training in early 1975 and went "on contract." I signed my sales contract and my compensation plan and was assigned a territory and a quota. My territory was everything west of Central Avenue (basically half of Phoenix) and all car dealers in the Phoenix area. My compensation plan was $10,000 per year in salary, plus commissions on everything I sold, I think 5% of gross sales. I spent considerable time with a calculator playing "what if" games on my potential commission earnings. To Madelin and me, it seemed like a fortune.

Burroughs had a good salesforce in those days. I still believe that it was one of the best in the industry. It was a young person's industry too. Just about everyone was under 40, white and male – not right, but that is the way it was.

Burroughs, IBM, and most of the other computer companies recruited mainly from colleges – especially for salespeople. They hired people that would fit into the company culture and then continually groomed them deeper into the company culture. We were all somewhat brainwashed. There was deep loyalty between the companies and the employees, something that would slowly erode over the next 20 years until it was pretty much gone completely in the 1990s.

Burroughs trained me well. They were determined that we understand the basics of selling, the technical features and benefits of our computers, and how they compared to the competition. In my rookie exuberance, I focused a lot on that. We were trained to talk about the features and benefits of the hardware, such as the number of peripherals our systems would support, the advantages of removable disks, the advantage of having a printing console, how much memory our systems had, and all sorts of things like that. But I would soon learn that those things did not matter much.

We sold primarily to small and medium-sized businesses including wholesale distributors, manufacturers, law firms, accounting firms, hospitals, construction companies, and car dealers. Most of the time, the decision-makers were the owners, the general managers, and the finance managers.

These people did not really want computers, nor did they understand them. They wanted solutions to their business problems. I heard someone say early in my career that nobody wants a drill, but everybody wants holes. That is so true.

Just about the time I was assigned my territory and started selling, IBM released the IBM System 32. It was aimed at the Burroughs B700 and other similar computer systems. Mainly though, it was designed to keep their IBM System 3 customers in the IBM family, and that it did.

But make no mistake, they wanted to take market share from Burroughs and the rest of the BUNCH too. IBM always believed that its fair market share was everything.

The IBM announcement was a big deal. Like most IBM announcements, it was billed as a revolutionary change to the computer industry and in many ways it was. IBM had now legitimized easy-to-use business computers that did not require a data processing department to operate. It was a formidable challenge to us, but it also brought many new opportunities. It seemed that overnight, everyone was interested in buying a new small business computer.

The Burroughs corporate marketing people were all over it. Almost immediately we had all sorts of competitive material showing up at the office. Burroughs launched a competitive sales campaign with the name, "They Shall Not Pass." We had a mandatory all-day training class on the System 32 that was held on a Saturday.

This was a corporate-mandated event held across the country and maybe around the globe. The district product manager and others flew in from Denver to conduct the session. It was intensive and very motivating.

Everyone was relieved to see that the IBM System 32 was not particularly impressive on a feature/function basis. I distinctly remember the presenter passionately explaining how bad the floppy disk backup process was on the IBM system. He exclaimed that it was the equivalent of trying to push a bowling ball through a drinking straw. I probably wrote that down to use later because I liked it so much. He was convinced that we could beat IBM every time on that feature alone.

On a feature-by-feature basis, the B700 would beat the System 32 every time, but IBM rarely tried to compete on a feature-by-feature basis. It was not a winning fight for them. They competed by creating confidence and trust in IBM and by creating fear, uncertainty, and doubt (FUD) in the mind of the customer toward everything and everybody else.

To IBM it was never about the computer. It was always about their market share, how big their local office was, and that General Motors or somebody had just bought 5,000 units of whatever it was that they were trying to sell. It was about the successful careers of IBM clients and how good they would look to their management by making an IBM decision. It was about how valuable new IBM skills would be to their customers' careers. They were experts at painting pictures of dread, doom, and gloom for anyone foolish enough to buy from anyone other than IBM.

They were also masters at making technical mediocrity sound like technical genius. They excelled at selling vaporware, although the term did not exist yet. Vaporware is basically something that exists only as a concept or idea.

Soon after the IBM announcement, I was making cold phone calls one day from a directory of Phoenix wholesalers looking for B700 leads. One of the calls I made was to Copeland Wholesale, a clothing wholesaler. After identifying myself and telling the receptionist why I was calling I was put through to Allen Miller, the owner and President of the company. I introduced myself and told him that I sold computer systems for Burroughs.

He was cordial and told me that they were interested in buying a new computer system to automate their inventory, billing, and accounts receivables and that they were already looking at other systems. I felt like I had hit the lottery again. He agreed to meet with me at his office a few days later. I had my first computer prospect.

I couldn't wait to tell Madelin and my other boss, Bob. They were both excited for me, but Bob more than Madelin.

A few days later, Bob and I met with Allen. It was a good meeting, and from all indications, they were going to buy something soon. Bob led the conversation since I didn't have a clue as to what I was doing. Bob asked a lot of questions about what they did, who they sold to, and that sort of thing. He asked Allen what the main objectives would be for the new system, what business problems they were trying to solve, and a few basic questions regarding transaction volumes. The answers determined whether or not the B700 would be the appropriate Burroughs system. And it was.

We set the next meeting to do a more detailed analysis of the business and to meet with some of Allen's key people. When we got in the car to go back to the office, Bob told me he thought I would soon have my first sale.

Listening to Bob at that first meeting was a great lesson. He barely mentioned Burroughs or the B700. Instead, he asked lots of questions and let Allen talk about his business. Too many salespeople talk too much.

The next meeting went well too. We learned more about the business and that they had already seen a demo of the IBM System 32 and one of the NCR systems. They gave us some invoices and a partial list of their products, including part numbers and descriptions. We scheduled an appointment for a B700 demonstration at our office for the next week or so. For the next week, all I did was prepare for the demo.

We would be demonstrating the Burroughs B700 with BMS application software for wholesalers. BMS stood for business management software or something similar. Over the next week, I learned some important things about the Burroughs product line I was selling.

The Burroughs B700 was not the most reliable system on the planet. And for consistency's sake, the BMS software was equally flakey. Further, the BMS corporate software support organization that I envisioned did not exist.

As I was preparing for the demo and trying to get all this stuff to work, Bob walked into the demo room and handed me a piece of paper with a name and phone number. He told me to call this person in Detroit (corporate headquarters) if I ran into any problems with the software. I had already "run into problems" so I immediately called this person. Someone else answered his phone and told me this person was out for the day. Naturally, I asked if I could speak to someone else in the B700 BMS software department to which he dryly responded, "He is the department." And so it was with Burroughs technical support.

I struggled with the demo software but was finally able to create files of inventory items with Copeland product descriptions, item numbers, prices, and actual names and addresses of Copeland customers. Once all of that was done, I started rehearsing the demo – over and over.

Sometimes, for unknown reasons, the B700 would go completely out to lunch. When this happened, a bell would ring, all the program key lights would light up and the print ball on the console would go all the way to the right, stop, and vibrate like a woodpecker pecking on a tree. There was nothing subtle about B700 demo death. All you could do was reboot by reloading firmware with a punched paper tape and start over.

Our B700 field engineers said it was the BMS software. The BMS software guy in Detroit said it was the B700. I would learn over the coming months and years that step one to a successful B700 demo would be the absence of demo death.

The demo day arrived, and I was nervous as crap but excited too. Allen Miller was there along with several of his key staff. Bob did the welcome and the Burroughs overview type stuff and I did the demo. It went great and much to my relief the B700 did not die. After we finished the demo, they lingered for a while for an informal and friendly discussion.

The thing I remember most is Allen bringing up the IBM System 32 again and the demo they had seen recently. He was not impressed with the System 32 or their demo but was quite impressed with ours. My new job was starting to seem easy.

Within a few weeks, and after several more sales calls, I got that call I had been waiting for. Allen asked if we could come by to discuss the last few issues. We were on our way with the contract in my Sears Samsonite briefcase.

Allen told us that they had decided to go with the B700, but that he wanted a 90-day trial period. Then, if everything went ok, he would sign the contract. I was completely over my head, so Bob led the conversation from there.

Bob told him we could do that with an order letter stipulating exactly what we had to do during this 90-day period to get the contract signed. Allen was fine with that.

Bob told him we would have the letter back to him by the end of the day for him to sign. It was the infamous Burroughs "order letter." On the drive back to the office, Bob told me how this would work.

Once we had the order letter from Allen, we would complete the ROPO and send it in. This was the internal document that was sent to corporate to place the order. I think ROPO stood for record of product order or something close to that. Yes, we were a computer company, but none of this was computerized back then. Then when Allen signed the contract in 90 days, we would put it in the customer file, which stayed in the office. He assured me it would be fine.

Allen signed the letter later that day and I had sold my first computer – sort of. As great fortune would have it, it all worked out fine. Allen signed the contract 90 days later and Copeland Wholesale went on to become a good Burroughs customer.

My first commission check for around $1,500 came soon. Madelin and I thought we were rich. We went to the mall and bought a bunch of stuff for our house – including a much-needed couch. It was so much fun.

The Burroughs L Series sales guys were expected to install the software, make any necessary code modifications, and sometimes write custom computer programs for the L Series product line. They were seller-doer pioneers. More on that later.

The same was expected of the B700 salespeople even though it was not even in the realm of practicality. Still, it was my responsibility to install the BMS software for Copeland and to make it work.

Within a week or so after we got the order (letter), Bob

walked over to my desk with a huge mainframe computer tape. He told me that it contained the source code for the BMS software for Copeland and that he had arranged for me to have time on the mainframe at night to compile the source code. I am sure I tried to act cool, but I clearly remember thinking, "Oh, shit."

This falls in the category of lucky being better than good. Just a few days later, a man named Hans Ager came to meet with Bill, our Branch Manager. Bob told me about the meeting and asked me to sit in. He said that Hans was a contract programmer but that nobody in the office knew him.

Hans was an impressive man, probably in his mid-to-late 30s with an equally impressive background in software development. He explained that he was opening a software development company in Scottsdale and wanted to specialize in Burroughs systems because of his Burroughs background.

We told him about our new order with Copeland and asked him if would be interested in implementing the BMS software. He said absolutely and quickly convinced us that he could do this with ease. The more he talked, the more I prayed, "Thank you, God."

Hans would go on to become a key figure in my early career. He was an important part of my success in the coming years, and I would work for his company some eight years later. He was a brilliant software developer and was, without a doubt, the best salesman I have ever known.

We agreed during the meeting that we would introduce Hans to Allen Miller at Copeland. The plan was to propose to change the BMS software price from $5,000 to $1,000 and to have Allen sign a separate $4,000 contract with Hans to do the software implementation. I was happy to forfeit $4,000 of my $40,000 sale in exchange for a workable plan for a successful implementation.

We soon introduced Hans to Allen to discuss the new plan and Allen enthusiastically agreed to it. To be honest, I would have outsourced the software implementation to a dead cat if I had had no other options.

Over the next two and a half years, I sold several B700s to other businesses in Phoenix that were similar to Copeland, all with Hans doing the software. But my real success would be with car dealerships. I never again had to worry about implementing software and, for the rest of my career, I never came close to writing code or compiling software.

Burroughs had several L Series installations in Phoenix-area car dealerships, so I started calling on them. The first prospect was a Volkswagen dealership in Scottsdale that had outgrown their L Series. Like everybody else, they were already looking at the IBM System 32.

I called Hans to see if he would be interested in developing a B700 auto dealer software package and to ask him if he had any experience with car dealers. With his customary cocky flair, he said, "Yes, I am, no, I don't, and it doesn't matter." He was right on all three.

I introduced Hans to the dealership, and they loved him.

Hans was a genius. I sat in awe as Hans talked with the management people at this dealership. It was as if he was the world's greatest authority on car dealership software. They would ask him how he handled this and that and he had a brilliant answer to everything. He never said no and never added the standard caveat that this feature "could be easily added." He just said yes, explained how it worked, and offered to demonstrate it.

After the meeting, I told Hans that I had no idea that his software did all that. He told me that it was just a simple matter of programming and that it would be fully functional by the time the demo came.

Soon, this became the easiest sale I had made yet. My new sales plan was to introduce Hans to every car dealership in Phoenix. Hans was hiring people and his company, Applied Data Systems (ADS), was opening a new office in Scottsdale.

From this point on, I focused on selling to car dealers but did sell a few other systems as well – always with ADS providing the software.

After a couple of months, we were able to use the VW dealership as a reference. They loved Hans and for good reason.

The car dealerships didn't really care about the B700 or why it was better than the System 32. They just wanted the ADS software and it happened to run on a Burroughs B700. It was that simple.

Basically, my job became getting appointments with car dealers and taking Hans with me. I sold a lot of B700s to car dealerships (and a heavy equipment dealership) in the Phoenix area over the next couple of years. I established myself as somewhat of the "Mr. B700" of the Mountain States District. But truth be told, Hans did most of the selling.

Over the coming years, ADS began to focus more on heavy equipment dealerships, rather than car dealerships. Hans eventually landed national endorsements of both John Deere and Caterpillar for dealership software. He spent much of his time flying all over the country in his plane growing his business.

But life was much more than just selling computers.

One of the two best days of my life was August 16, 1977.

Around 4 p.m. on August 15, my phone rang – it was Madelin. She hardly ever called me at work, so I knew it was something important.

She was working at SurgiCenter, an outpatient surgery and pain treatment facility in downtown Phoenix directly across the street from Good Samaritan Hospital. She was more than eight and a half months pregnant with our first child, Brian.

She told me that her water had broken and that we were about to have our baby. She had already talked with her doctor and was told to head to Good Samaritan. She calmly asked me to meet her at the special maternity admittance office and told me exactly how to find it. I was trying to stay calm too and was writing all this down. I asked her how she was going to get there, thinking that one of her nurse friends would be driving her. But no, she said she was going to walk – she insisted that it would be easier and quicker than driving across the street to the hospital parking garage.

It was extremely hot, probably around 110° plus or minus a few degrees. I told her I would come to get her, but she said no, that would be too complicated. I knew better than to argue, so I agreed to meet her as she wanted. It was only a 10-minute drive from my office.

I headed straight to the hospital and went to the maternity entrance as she described. It was on the west end of the hospital in the blazing afternoon sun. Madelin was standing there when I got there and told me the door was locked and that she couldn't get anyone to let her in. You had to ring the doorbell so they could buzz you in – but for some reason, we could not get in. We gave up and walked around the hospital to the main entrance. It was a very big hospital.

Soon we were all checked in and everything was good so far. She was not yet in labor, so she sent me home to get the stuff that she had already packed. We lived in North Scottsdale, which was about 45 minutes from the hospital. I did an exemplary job with all of this.

She was in labor when I got back. We had been to the child birthing classes and it was my job to be the coach – keeping time, reminding her how to breathe, massaging her back, and all that, but she was having none of it. I was sure she would still like the back rub, but she didn't like that either. She was very grumpy in labor.

Brian was born the next day at 5:10 a.m. He was a beautiful healthy baby boy, and Madelin did great. I remember thinking how lucky I was that the only baby I had ever seen that was actually beautiful was our baby – what were the chances?

I stayed with Madelin and Brian until around noon and then headed to the office to let everyone know the good news and to hand out cigars before heading home to take a shower, rest a little, and get the other stuff that Madelin needed.

When I got home, I turned on the tv and learned that Elvis Presley had died. It was the biggest news event since Nixon resigned. He died a few hours after Brian was born.

A few days later, we took Brian home. It was one of the happiest times of our lives.

7

THIS CHAPTER AIN'T NO JOKE

A couple of months after Brian was born, our new Branch Manager, George Burns, walked up to my desk one day and asked me if I could come into his office for a few minutes. George seemed somewhat old to us, probably in his mid-to-late 50s or so. He had been with Burroughs for most of his career and had an impressive background.

He had been brought in six months or so earlier from the Midwest, Indiana I think, to replace Bill Savage who had left the company. It was definitely not a promotion for George, so it seemed to all of us that he probably came to Phoenix to retire, which he eventually did. He was one of the nicest people you could ever want to meet.

I followed George into his office wondering if I had done something wrong. He closed the door, sat down behind his desk, and said, "Congratulations, you are the new Zone Sales Manager for Las Vegas." I couldn't believe it. It was one of the most exciting things that anyone had ever told me. Everything else he told me was a big blur, as I just could not believe what I was hearing. I didn't know that there was even an opening there, and I sure didn't know that I was under consideration.

That was the way it worked then. You got chosen and you went, it was like getting drafted, which was fine with me. George went on to tell me that I would be contacted by the relocation department about the move package, the starting date, the comp package, and all of that. He told me that the branch manager, Clay Vega, was waiting for me to call him and added that Clay had told him that he would not consider anyone but me. I felt so honored and so important.

Clay Vega was well known for his long history of success and his flamboyant management style. He had been there for a long time and was somewhat of a legend in the company, especially in the Mountain States District. He was known as a hard-driving, take-no-prisoners manager with somewhat of a wild-man, renegade style. Everybody had heard of a least one or two of the many Clay Vega stories.

Not wanting to tell Madelin the awesome news over the phone, I headed straight home. I knew she would be excited too, and she was, but mainly for me.

We were high-fiving, calling our parents, and discussing all the things we would have to do to get ready to move. We had been to Las Vegas a couple of times for the weekend, and we always had fun there. Moving there was our next big adventure.

Soon I was on a plane for the short flight to our new city. I would be there for a couple of days to meet Clay and my new team and to get acquainted with our new soon-to-be hometown.

Clay picked me up at the airport that morning and we

went straight to the office. It was the first time I had ever met him. He was engaging and friendly, but his flamboyant personality was easy to see.

On the drive to the office, he went out of his way to stroke my ego and to make me feel welcome. He repeated exactly what George had told me. He said that he always hand-picked his management team, that he had been watching my success in Phoenix for quite a while, and that he didn't even consider anyone else. He told me that I was perfect for this job and that he knew I would be very successful.

The discussion then turned to real business. He told me that I had a good team except for one guy that I would need to fire and replace. I asked him who that was and what the problem was. He told me that he was just not cut out to be a salesman but that he wanted me to figure out who it was. I interpreted that to be some sort of test, but maybe he was trying to be fair to the guy.

He went on to tell me that I would make my own hiring and firing decisions but then added in a matter-of-fact way, "but no blacks or women." It was quite shocking to hear that, but it would not be the only shocking thing that happened that day.

For much of the day, I was in the office hanging out with Clay and meeting people. Late that afternoon, Clay asked one of the sales guys to go next door to the convenience store to get a case of beer.

Soon, my new sales team, Clay, others that worked in the office, and I were convened in Clay's office drinking beer, telling stories, and having a good time. It was somewhat of an ad hoc welcome/beer party/meeting.

But the hours kept going and the beer kept flowing.

Somewhere later in the party/meeting, Clay began to talk about my predecessor, and I could tell that the sales guys respected him and clearly missed him.

He had resigned to take a position with another company and had moved. In those days, if you resigned you were often portrayed by management as having bad judgment and as having been a rascally sort of person. Some in Burroughs management never wanted to acknowledge that there might be a better career outside the company. Clay was one of them. He was fiercely loyal to Burroughs.

Clay soon began to build me up, telling the sales team how fortunate they were to have me as their new manager. He went on to tell them about all the things I had accomplished in Phoenix and how much they would learn from me. It was awkward for me, but everyone was nice, agreeing with Clay and telling me how excited they were that I was there.

But then Clay began to berate my predecessor and he just wouldn't stop. Finally, one of the sales guys, Billy McGee, had had enough. He defended his previous manager and reminded Clay of what a good job he had done. He told Clay that he did not deserve to be berated like this. A heated argument ensued.

Clay was sitting behind his desk and everyone could see that he was becoming angry. Then, he stood up, calmly took off his tie and his watch, and put them on his desk. He rolled up his shirt sleeves and walked over to Billy. It became very quiet. With his fists clenched, Clay got in Billy's face and said, "You can take the first swing, and then I'm going to deck your young ass." There are some things you never forget. This was one of those for me.

Billy, and the rest of us, tried to calm Clay down and to

defuse the situation. Thankfully, he sat back down and changed the topic as if nothing had happened. The party/meeting broke up and everyone but me went home. I went to my hotel. The next day it was like it never happened. The guys told me it was just Clay being Clay.

When I got to my hotel, I called Madelin to check on her and Brian and to tell her about my first day. I'm sure I said, "You ain't gonna believe this shit" somewhere early in our conversation.

In one day, Clay had exceeded my wildest expectations of his reputation. I could not help but wonder what I had gotten myself into. Still, I was excited about my new job and our new adventure, but I was in for a wild ride.

Madelin, Brian, and I moved to Las Vegas and were in our new home before Christmas. Madelin soon had a nursing job at one of the large hospitals, and I was consumed with my new job.

Las Vegas was much different than Phoenix. Phoenix had a laid-back, gentle-spirit type of vibe. It was surrounded by suburban communities with names like Paradise Valley and Carefree. For a big city, it was clean and well-kept.

Not so with Las Vegas. It had a hard-edge feel and was less friendly. Somehow, our new city just seemed somewhat sleazy to us. It was as if the city officials did not care what it looked like in the daytime. It was often strewn with trash that had blown up against chain-link fences along vacant lots. Las Vegas was much different then than it is now.

Most of our neighbors worked in the casino and hotel businesses – everything from card dealers to bartenders. The new wore off soon and it was no longer fun to go to the casinos.

But 1977 brought more than just a new baby, new jobs,

and a move for Madelin and me.

It was a time of change in the computer industry too. New upstart companies like Digital Equipment Corporation (DEC), Data General Corporation (DG), and others were making a real impact. DEC released the first of the 32-bit VAX line of computers that would revolutionize the small and midrange computer market. Soon, it would no longer be just a race between IBM and the BUNCH companies.

These were new, nimble, and innovative companies unburdened by decades of tradition and habit found in IBM and the BUNCH companies. It was the beginning of a culture change in the industry too.

IBM and the BUNCH were, in many ways, sales-and-marketing-led companies. Not so with DEC and the other "new" companies. They were founded and led by engineers with an engineering mission – to have the best computer technology in the world. Everything they designed was innovative – the microprocessors, the storage systems, the operating systems – everything. They believed that if you built a better mousetrap, the world would come, and for the next decade it did – especially to DEC.

At first, DEC, and DG to a lesser extent, were focusing more on technical and engineering markets and not so much on the business computer market. I think that is at least partially why the industry was caught somewhat flat-footed. Their midrange technology was elegant and astonishingly more advanced than that of IBM, Burroughs, and the other BUNCH companies.

Further, almost overnight, a new industry sprang up – resellers of DEC and DG computers. DEC and DG invested heavily in developing alternative sales channels not relying solely on internal salesforces. They recruited and established a large stable of third-party application software providers, hardware distributors, and turnkey system providers.

Within a few years, there would be a wealth of software available for these systems for practically every industry – far beyond what Burroughs could offer. Soon we were competing, not just with DEC and DG, but with a tidal wave of new channel companies that had embraced DEC and DG.

The B700 and the IBM System 32 were quickly becoming obsolete. These systems were single-task/single-user systems designed to do one thing at a time. You could print something, or you could run some sort of input job. You just could not do both at the same time. The new computers from DEC, DG, and others were designed to do multiple things at the same time – like mainframes could do. Many small and mid-sized businesses, and departments in larger corporations, were no longer content with the limited functionality of the B700, System 32, and other similar systems.

Enter the Burroughs B80 and B800 computer systems and the IBM System 34.

Everyone wanted multitasking systems with the ability to connect multiple workstations. IBM answered with the System 34 and Burroughs with the B800 and with the smaller B 80 (sort of). On paper, the B800 was superior to the System 34. The IBM System 34, however, worked mostly as advertised. This was somewhat of a role reversal as it was usually IBM that was selling vaporware.

A major advantage of the B800 was its theoretical ability to utilize a simplified version of Burroughs' flagship mainframe operating system, MCP. This was intended to provide the B800 with sophisticated multiprogramming capability and the ability to operate with multiple online workstations akin to the larger mainframes.

I was fortunate that the B80s and B800s were released just as I was starting my new job. Las Vegas had a large installed base of L Series machines and still had a lot of L Series prospects. The sales guys knew these machines well, and I didn't know much at all about them.

It was a new day though. The salesforce would transition to the new B80s and B800s during 1978. At the end of 1978, Burroughs adopted a theme and tagline for 1979, "Online in '79!" If you were not around then, that probably makes absolutely no sense.

None of us knew anything about these new systems, so I was at least on equal footing with my team from the new product perspective – but I was way ahead on B700 systems.

Clay was right. I had a good sales team and a big one compared to most zones. My team consisted of ten or so salespeople during 1978 and was one of the largest zones in the Mountain States District. In 1979, my zone would be split into two zones.

I was soon immersed in my job, making sales calls with my team, assisting in demos and proposals, meeting with unhappy customers, interviewing sales candidates, and doing all the things that zone managers did. In some ways, it was much different than my three years as a salesman but in other ways, it was much the same.

The most important thing I had to learn was how to work and get along with Clay. He was a successful and experienced executive, and I learned a lot from him even though we disagreed on some big issues and management styles. As I had learned on my first day, he had a fragile temperament that you did not want to provoke. For the most part, we got along fine, but we never really clicked and never became close friends. Our management styles and personalities were quite different.

One of my first missions was to determine who it was that Clay had said I would need to fire. After only a week or so, I was fairly sure he was a trainee that had been on board for less than a year. He had an accounting degree and I believe he had worked for a while as an accountant before joining Burroughs, though I'm not sure about that. In any case, he was young like everyone else, probably in his mid- to-late 20s.

My first objective was to try to help him. I met with him one on one several times to try to understand what he was doing and to figure out what I could do to help. He was quiet, shy, and lacked self-confidence and sales instincts. His primary assignment at the time was to sell Burroughs calculators. He was lost, overwhelmed, and needed direction.

We agreed that he needed to become more focused and to pick a specific vertical market and stick with it. We decided that he would focus on medical offices as they all used calculators in their business offices and were easy to identify from the yellow pages and other listings.

A few days later he stopped by my desk and excitedly told me that had successfully made an appointment with one of the larger dental offices. I was happy for him and told him to let me know how it went.

After a couple of weeks had gone by, I asked him about it. As they say, you can't make this stuff up.

"Hey, how did the sales call go with the dentist's office?" I asked.

"Not very good," he replied.

He went on to tell me what had happened. He had checked in with the receptionist explaining that he was with Burroughs and that he had an appointment. Soon, someone came out and escorted him into one of the examining rooms and told him to have a seat – the only seat was the dental chair. Then he made the mistake of sitting down. You can see where this is going. Someone came in, clipped on the bib, and told him the dentist would be right in.

He grinned embarrassingly and said, "The good news is I only needed a cleaning."

I had identified the person Clay had mentioned. Soon, he moved on, and I was looking for his replacement. My first conflict with Clay was straight ahead.

I started recruiting right away to fill the opening. Burroughs routinely recruited on college campuses, so I contacted two nearby universities, the University of Nevada at Las Vegas and Northern Arizona University. Interviewing on campus was a new thing for me, and I must admit it made me feel quite important.

I spent a day at each campus and found it to be boring, disappointing, and unsuccessful. I'm not sure I even invited anyone to a second interview. Most of the kids at least dressed appropriately but several came dressed in their hippie attire. I would tell them something about the potential job and get "cool, dude" or something similar as a response. For those, I just wanted to get the interview over as quickly as I could.

Thankfully, we also recruited off-campus as well, and I found several better candidates that way.

I received a random phone call from a young woman in

upstate New York. She told me that she had recently graduated from college with a degree in business and that she would soon be getting married and moving to Las Vegas where her fiancé lived. She knew about the Sales Trainee opening and said she wanted to interview for it during her upcoming trip to visit her fiancé. I was impressed with her confidence and tenacity, so we scheduled a time for her to come in for an interview.

Within a week or so she came in for the interview. She was very pretty (ok, I noticed) and equally likable. Her resume was impressive, and she said all the right things. She was the best candidate by far and I was confident she could be successful. I was less confident, however, that Clay would agree to let me hire her. He had made his opinion crystal clear on this issue, but I was up to the challenge.

I went into his office to tell him that I wanted to hire Denise. I told him that I knew how he felt about it, but I felt she deserved a chance. To my surprise, he said, "Ok, it's your zone" and then added, "Just remember, I told you so."

I was very relieved and confident that I was doing the right thing. I called Denise to tell her that she would be getting the offer letter soon. We agreed on a start date and that was it.

Her first day was to be on a Monday a month or so later. We had spoken a few times by phone, but I had not seen her since the interview.

I was excited about her first day and made sure to be in the office when she arrived. She didn't show up. I knew something must be very wrong, but I didn't know what. Around 10 a.m., she called me – crying. She told me that she and her fiancé had broken off the engagement and that she would not be moving after all. It had just happened – she was very apologetic and quite upset. I felt bad for her but also felt that she had let me down. It was as if Clay had some sort of crystal ball. I told her that I was sorry, wished her the best, and prepared to go tell Clay.

It was a short discussion. Clay did not lecture me or ask for a lot of detail. Barely looking up, he just said, "I told you. Now go hire someone that will show up." I left his office feeling humiliated, but not defeated.

Filling this opening was a high priority for me – now even more than before. Along with all the other things I had to do, I continued the search for Joe's replacement.

One day shortly after all this happened, the receptionist came back to my desk and told me that someone was in the lobby regarding the open position. I wasn't expecting anyone but went out to the lobby to see if it may be someone that could be a potential candidate. I believe that the whole Las Vegas experience was one giant test that God put before me.

Yes, it was another professional-looking and attractive young woman. She introduced herself and apologized for dropping in without an appointment. She handed me her resume and asked if she could make an appointment for an interview. She told me that she had been a high school teacher in New York and that she just could not do that anymore. She had moved to Las Vegas looking for a new start. Her name was Shelley Jones.

Within a few days, she was back in the office for an interview. She didn't exactly fit the mold. She was somewhat older than most trainees, she did not major in business, and she was, well, a woman. She had been teaching for seven or eight years, so she was probably 30 or so. She was single, businesslike, polite, and had a certain New York toughness about her. She was quite assertive, almost overly so, and made it clear that she was determined to build a new career in sales. She had done her homework on Burroughs.

Within a few weeks, I had several candidates, including her. I knew that this may be my death knell with Clay, but it was an easy decision to choose Shelley.

I called her and asked her to come into the office again. We had a very direct discussion with me telling her things that no one would say today. As diplomatically as I could, I told her that Clay was not very "progressive" with certain things. It was 1978 – she knew what I meant.

With a bit of trepidation, I told her that I needed to know for sure that she would take the job if it were offered to her and told her why I needed to know. She assured me that she would and that she could handle Clay. I had a sense that she could indeed.

I had to find the right time to tell Clay, so I waited until I knew he was in good mood, which took a few days. I tried to be somewhat funny about it. I started the conversation with, "You aren't going to believe this." He laughed and said something about me being a slow learner. He reluctantly said ok but told me that this time he wanted to interview her. This was a recipe for disaster, but I had no choice. He gave me the day and time to have her come in.

I called Shelley to tell her and to prepare her for the impending shit storm. She was happy that Clay wanted to talk with her and told me not to worry.

Shelley and I talked several times before the Clay interview, strategizing about how to handle some of the things that I knew Clay would surely say. Although I did not tell Shelley, I had decided that it would be a successful interview if it did not end in a lawsuit.

The day of the interview came, and Shelley and I were ready. She was to be in the office late in the afternoon, 4 p.m. or something like that. Then the wheels started to come off. Clay walked up to me in the middle of the day and told me to have Shelley meet us at a local bar, instead of the office. This was certain disaster. I called her to change the location.

Sometime shortly after lunch, Clay came by my desk and said, "Let's go." I reminded him that she would not be there until 4. "Yeah, I know. We can have a few drinks before she gets there," he said. This was a horrific idea, but I couldn't stop it.

We got there and started drinking. I did the best I could to pace myself knowing that Clay would not. By the time Shelley arrived, we had both drank too much, especially Clay.

The meeting (and I guess you could call it an interview) started ok but it soon derailed. Clay had a reputation for being "direct" and he took pride in it. He was even more "direct" when he was drinking.

His first challenge to her was around toughness. He told her that we were in a tough business, that our competitors were tough, our clients were tough, and that he was tough. He asked her if she thought she could handle this sort of job.

I will never forget her response. It was basically this: "Mr. Vega, there is nothing about this job, or you, that intimidates me in the least." She then asked him if he had ever been backed up against a wall with a knife at his throat.

Clay said, "No."

"Well, I have," she said.

She told him that she had worked as a teacher in one the

toughest and most dangerous high schools in New York City and that her life had been threatened several times, once while being held against a wall with a knife to her throat. Calmly, but sternly, she told him that she could handle his tough job, his tough customers, his tough competitors, and him. I was awestruck by her response, and Clay was too. I thought he was going to offer her the job right then.

I thought the worst was over, but no, Clay reloaded and came at her one more time. This is uncomfortable to even talk about. He asked her if she had met any of the sales guys in the office, to which she replied, "Yes." He then boastfully told her that he hired "studs." He leaned in toward her, looked directly into her eyes, and asked if she would be f-ing the sales guys. This was a shocking and embarrassing question, even for Clay.

Most women would have walked out and headed to the nearest law office, but not Shelley. She calmly replied that this was none of his business and that if she did, he would never know about it. Clay loved that response. He told her that he liked her and that he knew she would do great. I had another "Clay story" for Madelin.

A few weeks later, this large Burroughs Branch had its first woman sales representative. From then on, Clay treated Shelly just like "one of the guys," no better and no worse. As for me, although he never said so, I think he appreciated that I did what I thought was right and hired the best person for the job. When I left Las Vegas, she was doing great. I never heard from her again, but I am sure she had a successful career.

Except for the one time that Clay fired me for a few hours, we worked well together, despite our many differences. For all his eccentricities, Clay was a brilliant businessman in many ways, and no one could argue with his success.

One night, as happened quite often, we were all in the office after work drinking beer, reorganizing the company, and solving all the world's problems. Somehow, Clay and I got into an argument about something. I knew better. It was never a good idea to argue with Clay, especially in a beer meeting. The argument became quite heated, and Clay finally exploded and told me that I was fired. I finished my beer, calmly I should add, and then went home to tell Madelin what had happened. I told her not to worry and that it would blow over.

She knew Clay and was less worried than I was. Sure enough, Clay called me at home within 30 minutes or so. He laughed it off, changed the subject, and said, "See you in the morning."

I was very ambitious, loyal to Burroughs, and committed to moving up the Burroughs ladder. Typically, the next rung for a Zone Sales Manager would be a Product Manager position on the district staff, which was in Denver, or a sales training position in one of the corporate training centers. Either would have been fine with Madelin and me, although we would have preferred Denver. Neither happened, but Clay had submitted the paperwork for my promotion to the training staff in Pasadena shortly before I resigned.

8

MIAMI VICE

In the spring of 1979, one of my previous managers from Phoenix called me. He had left Burroughs and called to tell me that Sperry Univac was looking for a Branch Sales Manager, I believe in Denver, but for certain somewhere in the west. He went on to tell me that Sperry was a great company and that he wanted to recommend me for the job. I told him I would talk it over with Madelin and call him the next day.

As much as Madelin and I loved living in the west (except Las Vegas), we had talked about trying to get back east somehow. Our families were in Mississippi and Florida and our son, Brian, would be growing up mostly without them if we stayed out west. We felt that anything east of the Mississippi river would be better in that regard. We decided not to change companies unless it would be an opportunity to move back east.

I called my friend back the next day and told him that if I were to leave Burroughs it would have to be for something back east. He gave the name and number of the HR Director for Sperry's Business Systems Marketing (BSM) Division at their world headquarters in Blue Bell, Pennsylvania, and said that I should call him because they were expanding all over the country. I called the next day and that call set in motion our move to Florida and a big change in my career.

The HR Director was Jurrance Peace. He was nice and interested in the possibility of bringing me into Sperry. I explained to him what I was doing at Burroughs and why I would consider leaving for a move back east.

He immediately asked me if I would consider a sales management position in Miami. He told me that there was an opening there for a Branch Sales Manager in the BSM division and that it was one of their most successful branches. I agreed to send my resume so that he could share it with the eastern regional management team. He told me he would be back in touch within a few days. I was excited about the possibility and couldn't wait to tell Madelin.

He soon called me back to tell me that they wanted me to come to their world headquarters in Blue Bell to meet with HR and some of the eastern regional managers. If that went well, I would then go to Miami to interview with the Florida BSM manager.

He gave me all the logistics: how they would send the airline tickets, which hotel I would stay in, the agenda for my day in Blue Bell, and the arrangement for a limo to pick me up at the Philadelphia airport to take me to the World Headquarters building. I think he had me with the limo.

This was all quite overwhelming to me, and it was the first sign of the significant cultural difference between Burroughs and Sperry. In a similar situation, Burroughs would have told me to rent a compact car and stay at the nearest Holiday Inn.

Everything went according to plan and soon I was in Blue Bell for the day of interviews.

Little did I know that within a few years, Madelin and I would live within a few miles of the Blue Bell headquarters. Nor did I know that Blue Bell would eventually become the world headquarters for Unisys, which would be created by the merger of Burroughs and Sperry in 1986, and that I would spend 23 years working for Unisys.

My day in Blue Bell went well. They all liked me, and I liked them and everything they told me about Sperry Univac. At the end of the day, Jurrance told me I should plan on flying to Miami soon to interview with the Florida management team. Within a week I was in Miami for more interviews.

I bought a new suit, a tie, and a white shirt for the Miami interview. I made the short flight to Los Angeles for a connecting redeye to Miami. I landed in Miami early the next morning, put on my new suit in the airport, and headed to the Sperry office in the Miami Koger Office Park.

The culture change was continuing. The Sperry Univac office was quite upscale compared to what I was used to. It took up most of the first floor in one of the large, mostly glass buildings in the Koger Center. The reception area was very nice with a fancy Sperry Univac sign accented by spotlights behind the reception desk. There were lots of large live plants too, something Burroughs did not bother with.

The Branch Manager, Ron Archer, came out to escort me back. Before going to his office, he gave me a quick tour of the facility including the demo room, the field engineering area, and the large area of the building that housed the America's division, which was the Sperry Univac mainframe operation. Everything was more formal than I was accustomed to. The furniture, including filing cabinets and desks, was modern, and everyone that did not have a walled office had modern office cubicles.

Ron and I immediately hit it off and it was soon clear to me that they would be offering me the job.

The cultural difference extended well beyond cosmetics. Something that really impressed me was their commitment to customer support. I was introduced to the manager of customer support, Alex Cabrerra, and he introduced me to several members of his team. They were responsible for all software implementations and customer support for the software. They were referred to as CSRs – customer support representatives. Sperry had as many CSRs as they did sales representatives. At Burroughs, we had one similar position for the entire office.

They gave me a quick demo of the Sperry Univac BC/7. The BC/7 computer line included several models but was the only product line of the BSM Division. Like several competitive systems, including the IBM System 34, the Burroughs B800, and the DEC VAX computers, the BC/7 had been released in 1977. This is where the "7" in BC/7 came from. BC stood for "business computer." Sperry had also developed their own business applications such as order entry/inventory, accounts receivable, payroll, and so forth. The BC/7 was designed to handle multiple workstations and it worked well. The Florida office had a huge installed base of these systems – maybe a hundred or more.

I was offered the job, accepted it, and soon Madelin, Brian, and I were living in Miami. We bought a house in Miami Lakes, not far from the Koger Center.

Madelin went to work at Jackson Memorial Hospital. Brian

was less than two years old in daycare listening to Spanish in the daytime and English at home. The poor little guy had to be greatly confused trying to learn to talk.

The differences between Sperry and Burroughs became more apparent with each passing day. I was amazed, and still am, at how two companies in the same industry could be so different. These differences would become more than just interesting observations when the two companies would merge some seven years later.

It struck me that Sperry seemed like a computer company for grown-ups. This was my observation based on having worked in two Burroughs locations and one Sperry location. Maybe it would have been different with a larger sample size – but I don't think so.

Most everyone in the Miami office was older than the people I worked with at Burroughs. Like me, they all had come from other computer companies or had been with Sperry for quite a while. There were no trainees and few, if any, people under 30 in the office – except for me. Things just seemed to be more serious. As Pete Griffen had told me during my first Burroughs interview, Burroughs worked hard and played hard. Sperry just worked hard, at least in Miami.

When I interviewed in Blue Bell, they had explained to me that the Business Systems Marketing division was a new and separate division of Sperry Univac dedicated to the small business marketplace and the BC/7 product line. It seemed perfect for me, fit my background well, and seemed like a great way to begin my career with Sperry. The small business computer marketplace was booming and with no end in sight. However, I did not realize how insignificant the BSM division was. Sperry Univac was a mainframe company and the BSM division was basically an experiment.

I had another good sales team. Unlike my Burroughs team in Las Vegas, the Sperry guys were older – not old, just older. The Miami office had been successful before I got there and was one of the most successful in the country. My time there was no different.

One of the most memorable deals we won while I was there was with a wholesale distribution company in downtown Miami that had a Burroughs B800.

One of the sales guys came into my office and told me that he had a prospect that was unhappy with Burroughs and their B800 – so unhappy, in fact, that they were considering getting rid of it and starting over with something else. He asked me to go with him on his next sales call which, of course, I was happy to do. Besides, that was my job. I was somewhat suspicious that this company had been using us to threaten Burroughs to force them to fix whatever their problem was, but that turned out not to be the case.

We met with the President of the company just a few days before hurricane David was forecast to hit south Florida. He was a nice guy but was angry with Burroughs. They had been using the new B800 for a while but were not using any of the online terminals (which everyone generically called CRTs). They were still in the boxes. He said that Burroughs had repeatedly ignored their requests to get them installed and functioning. I was quite sure I knew what the problem was.

He took us to the computer room to let us look at the B800 and to introduce us to the computer operator. I knew the B700 well enough to recognize the system commands. I took one look, and it was just as I had suspected. I told him that it appeared to me that it was operating with the older SCP-based operating system. If I was correct, his B800, in essence, thought it was a B700. He was not a happy camper.

We then went back into his office where he told us that his next phone call would be to Burroughs adding that, if what I had told him proved to be true, we had just made a sale for a Sperry BC/7 computer. To give Burroughs the benefit of the doubt, I told him that sometimes Burroughs and the customer would agree to install the B800 with the simpler and older SCP-based operating system to later upgrade it to the newer MCP-based one. He assured us that this was not the case.

Hurricane David brushed Miami and went inland further up the coast, but it had come close enough to shut down everything in Miami for a couple of days. True to his word, within a few weeks after David passed by, we had an order for a BC/7 system to replace the B800.

And here is one for the "Miami Vice" department.

A man came into the office one day with no appointment and asked to speak with a salesman about buying a computer – this NEVER happened. The receptionist came back and to get one of the sales guys to talk with him. It wasn't long until the sales rep came into my office telling me that this man wanted to buy one of the demo computers, pay cash and take it with him. It sounded like the flakiest deal (and man) ever.

For starters, taking a BC/7 computer with you would be the equivalent of taking a washing machine, clothes dryer, refrigerator, and range home with you from Lowes. On top of that, these were not "do it yourself" computer systems –they required field engineers to set them up and complex software to make them do anything useful. And finally, the price tag started somewhere around $40,000. I joined them in the demo room anyway.

Sure enough, the man told me that he wanted to buy the demo system, that he would arrange for someone with a truck to come to get it later in the day, and that he would pay in full by cashier's check. The number one rule in selling is that when you have made the sale – stop selling. However, we violated this rule trying to ask a few basic questions, like, "What sort of business do you have?" That went nowhere. Soon we had the Finance and Accounting Manager in the demo room, which never happened either.

After several phone calls to corporate to determine how, and if, we could make such a sale, we had the approval. We prepared the contracts while he was gone to get the cashier's check. We sold him the system, he paid for it, took delivery later that day in a rental truck, and we never heard from him again.

When you sell something that you put little or no effort into, it is called a "bluebird." This was the ultimate "bluebird." It would have been an excellent premise for an episode of Miami Vice.

Madelin and I were both working harder than ever. I was very busy at the office, and Madelin had a demanding job at Jackson Memorial. On top of that, Brian turned two that year, and we were doing our best to be good parents despite our demanding jobs. We were yuppies.

I loved my new job, the Sperry culture, and the people I worked with, including my boss, Ron Archer. It was a complete turnaround from my time in Las Vegas. Everything at Sperry was regimented, businesslike, and "normal." No beer meetings.

After the close of each quarter, the management team

(the branch manager, customer support manager, finance manager, and I) would go to our world headquarters in Blue Bell for the quarterly business review (QBR). This was an expensive and relatively useless exercise in slicing and dicing the numbers in every possible way that they could be sliced and diced. It was also a sales forecast and outlook for the rest of the year. It had the formality and seriousness of a court proceeding and was not for the faint of heart. Fortunately, we always had good news to report and a good sales forecast. Further, the finance manager did most of the prep work. It also gave me the opportunity to get to know some of the corporate executives on a first-name basis and to begin to establish a good reputation with them. This would soon serve me well.

Within a year or so after I arrived in Miami, the Business Systems Marketing division and the BC/7 product line were discontinued by the company.

Sperry Univac made a strategic decision that the small business computer marketplace was not for them – and they were probably right. While the BC/7 line of computers had been competitive, the new division had not met the expectations of the company. Further, the BC/7 was approaching the end of its life and falling behind from a technology perspective.

Most of the people in the Miami office worked in the mainframe computer division – called the America's Division. There were probably only 30 or so people in the Miami office in the BSM division.

Fortunately, I had come to know some of the America's people, including Carl Straw, the Florida manager. He was a true professional in every way, well-respected and liked by everyone.

Sperry called concurrent all-hands-on-deck meetings across the company – all of this had been a well-kept secret. There were people at our meeting that none of us knew, including some corporate HR folks – always a bad sign. Carl Straw, the Florida Manager of America's Division was there, and it was obvious to all that something big was up. This is when the announcement was made that our division was being discontinued by the company.

Corporate downsizings, mergers, and massive layoffs would become commonplace in the tech industry in the years and decades to come, but this was quite unusual at the time and it was certainly new to me.

The formal part of the meeting did not take long at all. The corporate people put a positive spin on it, as best as they could, and assured everyone that it was the company's goal was to find everyone new positions elsewhere in the company. The HR people were there to immediately begin meeting with everyone individually.

After the meeting was over, Carl Straw called me into his office. I knew Carl, but not very well. He was reassuring and told me that I would be "just fine" and that this would probably prove to be a good thing for my career. He began by telling me that he would love to have me work for him in Miami as an Account Executive in the mainframe group if that was something I wanted to do. Looking back, that was probably a much better option than I realized at the time – but my ego and ambition got in the way.

He made a point to let me know that the company was assuming that I would want to stay in management and that they were committed to making that happen if I would be willing to relocate, probably to somewhere in the northeast.

He explained that there were several openings that would be good options and that HR would discuss this further with me. I told Carl that I would talk it over with Madelin and would let him know the next day.

I was not looking forward to this discussion with Madelin. There would be no high-fiving this time.

Much to my relief, the discussion went better than I feared. As always, she was supportive and at least open to the possibility of moving again, although certainly not excited about it. The big question, of course, was "Move where?" We decided to at least take the next step to find out.

The next day I met with Carl and told him that we would consider moving depending on the location and the specifics of the job. Carl introduced me to the HR person and, before the end of the day, I was told that there were three locations that had openings for a Branch Sales Manager. The openings were in Montclair, New Jersey; Detroit, Michigan; and Philadelphia, Pennsylvania, none of which sounded particularly appealing.

Chuck Cliburn

9

THE REINVENTION OF THE 1980S

After Madelin and I talked it over, we decided to stay on the management track and that Philadelphia sounded like the "least worst" of the three options. The Philadelphia Branch was located in the same building as the Eastern Operations headquarters in Wayne, a suburb north of Philadelphia near Sperry Univac's world headquarters.

Having been to Blue Bell several times, I knew that the northern suburbs were quite nice. Additionally, working near world headquarters and in the same building with Eastern Operations were definite pluses that should be good for my career. I was soon on a plane to Philadelphia to interview for the position.

I interviewed with the Philadelphia Branch Manager, the Vice President of Eastern Operations, and several others. That was one of the disadvantages of being in the same building with Eastern Ops – everyone thought they should get to vote on everything. But it all went fine, and by the end of the day, I was offered the job. I called Madelin to get her final ok and then accepted the offer.

Miami had seemed quite formal and serious to me compared to my years at Burroughs – but Philadelphia was off the "serious scale." It was all business. Everyone was somewhat polite, but most were not particularly friendly. It was just the Philly/Jersey way and something that Madelin and I never got used to.

My new boss, John Smith, was the most intense manager I have ever known. He was not one for small talk and not one for excessive friendliness. One of the managers I interviewed with warned me that John was "wound up a little tight" and added, "Just roll with the punches and you will be fine."

I flew back to Miami, and Madelin and I started the moving process – again. I made all the arrangements with Sperry's relocation department and, after a week back home in Miami to wrap things up, I was on my way to Philly to start my new job. I would return to Miami once more to get Madelin and Brian and to make the long drive to Philadelphia.

It was now 1980 – the first year of a decade of reinvention for the IT industry.

From my view, the '80s were to information technology what the '60s were to music. If you were unfortunate enough to miss the 1960s, be advised that the music of the early '60s was absolutely horrific but the music of the mid and late '60s was absolutely awesome.

In 1960, the top 40 included songs like *Teen Angel* by Mark Dinning, *I'm Sorry* by Brenda Lee, and *Puppy Love* by Paul Anka. From my view, it all sucked. In 1969, the top 40 included songs like *Magic Carpet Ride* by Steppenwolf, *Honky Tonk Women* by The Rolling Stones, and *Hey Jude* by The Beatles – it was awesome. Of course, this all started when The

Selling Information Technology

Beatles came to America in February of 1964. Today, the music of the mid and late '60s lives on and is played every day on classic rock stations around the world and by cover bands (including my band, The RockitZ) in bars and music venues in everyone's hometown.

The 1980s brought similar changes to the information technology industry.

To put this in perspective, when I started my job in Philadelphia in 1980, I was working for one of the world's largest computer companies in a large Sperry Univac facility. But consider this: there was no internet, no cell phones, no PCs, no email, no PowerPoint slides, and no Excel.

When you came into the office, you would stop at the secretary's desk and pick up your stack of "While you were out" paper messages. The secretary would answer your phone and handwrite the message: "Call Jim a.s.a.p. at 885-3455."

When you were at off-site company meetings, there would be a stampede of people during every break rushing to the payphones so that your secretary could read you all your "While you were out" handwritten messages.

There were no multi-party conference calls either. Outboard phone speakers were all the rage so that multiple people could gather in a conference room or office to participate on the same call. We were still using electric typewriters, mostly IBM Selectrics. And 35mm slide projectors were used for formal presentations. We called them sleep machines. For less formal presentations, overhead projectors were used with transparent copies of typed documents. A stampede of technology change was on the way.

There was a significant cultural shift during the '80s as well. At the turn of the decade, there was still loyalty between employers and employees – a genuine all-for-one and one-for-all team spirit. In general, companies viewed their employees as assets. We all felt valued by our company and the idea of routine and massive layoffs was not on anyone's radar. By the end of the decade, it would be much different and by the turn of the century, company/employee loyalty would be only a distant memory for most IT firms.

It was also a decade of social change in the workplace. In 1980, the computer industry was still pretty much an industry of young white men. It was rare to work with anyone over 50 in 1980.

There was not a lot of social or political correctness going on back then either. With some exceptions, women worked mostly as receptionists, typists, secretaries, and clerks. There were even fewer blacks working for IT companies. I do not recall working with any black people until the mid-'80s.

Sadly, crude and demeaning jokes and commentary toward women were all too common in the workplace in the '70s and to a lesser extent, in the '80s. The "dumb blonde" jokes were mild compared to other sexist jokes and other things that happened. Bad behavior that was commonplace in those days would result in immediate firings and lawsuits today. I could recite many embarrassing examples, but here is one of the milder ones.

When Burroughs released the B800 in the late '70s, the District Product Manager came to our office to lead the product announcement meeting. This announcement included a 35mm-slide presentation consisting mainly of pictures of the new system and its components. Each slide included beautiful young women clad in the skimpiest of bikinis. When the first slide came up, the presenter said, "Now

Selling Information Technology

that I have your attention." Everyone laughed. It seemed like no big deal.

I'm sure these were "rogue" slides, not the official corporate slides, but they were used in the Phoenix office and, I assume, in other offices across our district.

Wives were not invited to the company reward trips back then – at least not at Burroughs and Sperry. The Burroughs trips were called "Legion of Honor" and the Sperry trips were called "Go Club" or just "Club." These were company trips awarded each year to the top sales achievers. The Burroughs trips were usually regional events, but the Sperry trips were typically company-wide events in Hawaii or someplace outside the country.

These events could best be described as frat parties for men old enough to know better. In my view, the Legion of Honor was hardly "honorable" nor was Go Club. I could go on, but it's better that I don't. This began to change for the better during the '80s.

Everyone smoked in 1980, even Madelin. People smoked in their offices, in the conference rooms, in their cars, in restaurants, and in their homes. I guess we were all used to it. As a non-smoker, it didn't bother me much. By the end of the '80s, smoking in the workplace was at least frowned upon and it was banned completely in the early '90s by most companies and governments. Thank goodness, most people finally quit smoking, including Madelin.

So back to our move. The Sperry relocation department put us in a corporate apartment in Norristown until we could find and buy a new house. Sperry had so many people moving in and out of the Philly area, and others on temporary assignments, that they contracted with an apartment complex for Sperry people that needed temporary housing.

The apartments were furnished so this made the move somewhat simpler than normal. The company paid for the apartment and the storage of our furniture while we were there. All we had seen were pictures of the apartment complex in brochures, so we were moving in "sight unseen." They assured me that these were nice apartments in a nice area. Unfortunately, I had never been to Norristown and nice, I guess, was a relative term.

Madelin was not particularly enthusiastic about this move and neither was I. Having been to Philly several times, I did my best to prepare her for what I knew was coming and what she was envisioning – an old industrial city with lots of smokestacks and storage tanks. I assured her, however, that once we got to the northern suburbs it would be much different and that she would like it.

We drove from Miami to Philadelphia with our 2-year-old son in our Datsun 200sx. It was a long drive. Madelin's Toyota was in the moving van.

We had been driving all day when we finally arrived in Philadelphia. We had no problem getting to Norristown but finding the apartment complex was a different story. I kept telling Madelin that it was going to get better soon. But it kept getting worse. We had written directions to the apartment complex along with a map, but we were seriously lost. I managed, however, to find all the "not so good" sections of Norristown.

We persevered and finally arrived at the apartment complex. It turned out to be a decent section of town, but "nice" was probably a little generous. I was afraid that Madelin would not get out of the car, but she did. We were very tired, it was very cold, and she was not very happy.

Selling Information Technology

After a few months, we bought a new house in the community of Eagleville. It was a small development of new homes built on what had previously been a farm. The entire development was on a steep hill – so steep, in fact, that our driveway was put in on a diagonal across the front yard. Our house was downhill from the street. Our next-door neighbor, who also worked for Sperry, told us when we moved in to make sure we parked on the street before snowstorms, otherwise we would never get out of the driveway. That proved to be excellent advice.

I was soon at work, meeting people and getting into the routine of my new job. Some of the larger Sperry offices, Philadelphia being one, had dedicated sales and support teams for the data entry and distributed systems product lines. I was the new Branch Sales Manager for this group. My group sold specialized computers designed for high volumes of data input.

Data entry specialists would key in data via online workstations so that the data could be collected, edited, and stored on disk drives. Large batch files would then be transferred to mainframe computers via magnetic tapes. Data entry rooms were typically very large, often with hundreds of people keying in data from various hard copy documents. Our customers were mainly large insurance companies and banks, but other industries used these systems as well. One of our customers was Binney and Smith, the manufacturer of Crayola crayons.

Madelin went to work for a local hospital, and we were soon in the routine of our new life in Philly. We liked our neighbors and became friends with them but, other than that, we never seemed to fit in very well. My observation of the office during my interviews (polite, but not particularly friendly) turned out to be quite accurate. I rarely had lunch with anyone from the office. Most of the people in the office were from that area and had worked with each other and for Sperry for a long time. It was quite "cliquish." None of this bothered me much.

I had known Madelin since we were kids and for the first time in all those years, I could tell that she was not happy. She never complained, but this move had really affected her. She was quiet, not herself, and on the edge of depression.

We tried to make the best of it. We went out to eat a lot. and there were plenty of nice restaurants in the area. Most of them were in old houses all claiming that George Washington, or someone similar, once stayed there. We would joke about it. "George Washington peed in that bathroom," we would say.

On weekends we often went to antique shops and bought antique furniture, some of which we still have. Madelin did a lot of things with Brian too, including enrolling him in swimming classes in the middle of winter, which I thought was a little strange but wouldn't dare say so.

My boss was not a personable sort of guy. I never heard him ask anyone if they saw the game, how the family was doing, or what their plans would be for the weekend. He was all business and not particularly pleasant about that either. I don't think anyone liked him much, but no one, except me, seemed to have friendliness as an expectation. His style and personality were not that unusual for the Philly area. He would not have been a good fit for Atlanta, Tallahassee, or Jacksonville but, for Philadelphia, I guess he was fine.

I must say, though, that despite his lack of warmth, he always treated me well.

Just about everyone in the Sperry building was older

than I was, but that did not take much as I was still in my late 20s. One of the guys who worked for me was old enough to be my dad – probably around 60 or so. He was a good sport about it but would occasionally remind me that he was selling computers when I was still in grade school. Abacus computers, I presumed.

My group was somewhat the "F Troop" of the Philadelphia Branch. My sales team was a newly created group, created just before I got there.

To create the new team, the other sales managers in the office had to transfer some of their sales reps to the new group. Obviously, they did not give up their superstars. I did have one superstar though. Gren Foote had been successfully selling these systems for years and was great at it. And he was a nice guy too.

One of the guys that I admired a lot was a mainframe Account Executive that did not work for me. His name was Fran Ashburn. Fran was older than most of us, probably around 50. He was always nice to me, especially compared to most everyone else. I viewed him as somewhat of the elder statesman of the office. Fran would be the center of a significant incident in my life some 20 years later.

The upcoming year was a difficult year for me and my group. We were doing a lot of good things and had a good pipeline, but the orders were just not coming in. I felt a lot of pressure, but it was coming from me, not from my boss or anyone above him.

Somewhere around the beginning of the fourth quarter, which was January of 1981, it became all too clear that my group might not make our numbers. That was a bad thing back then and it still is.

I stayed late one day waiting for everyone to leave so I

could have a discussion with John about this without being interrupted. After everyone had left, I went into his office and asked him if he had a few minutes. I told him that my team was working hard and that, while we still could make our numbers, the reality was that we might not. I asked him for advice and if there was anything that he felt I should be doing differently.

Barely looking up from his desk, he told me that the darkest hour was just before dawn and to keep doing what I was doing. After saying that, he stood up and reached for his coat, signaling the end of the meeting. That was it.

That was as good as it could get coming from him. I left the office feeling better and knowing, for the time being at least, that I had his confidence and support and that I was not about to be fired.

Soon, the deals started closing – one big deal after another. Everyone on my team was closing the deals we had worked on all year. Our year ended on March 31. My team had made our numbers after all.

It was snowing that day. After work, we had a celebration dinner and party at one of the nearby hotel bars. Everyone was there including the secretaries, the finance people, the order entry people – everybody. It was in the middle of the disco era and everybody was dancing and having a good time. It was quite the celebration.

Madelin and I had already decided that, at the end of the year, I would ask for a transfer back to Southern Operations. We just could not stay there any longer.

The next day I went back into John's office to ask for a transfer. This was a highly unusual request, and I knew it.

I told him that I greatly appreciated working for him, but

Selling Information Technology

that Madelin and I wanted to move back to the south and that I would appreciate his help in making that happen. He told me that they wanted me to take a new management position in the Harrisburg branch. It would be a promotion and I would be responsible for the entire product line, including mainframes.

John told me the rest of the story that he could not tell me in January. He said the VP of Eastern Ops had wanted to create the new Harrisburg management position but that corporate would not approve the additional management slot unless he eliminated an existing slot. The slot he wanted to eliminate was mine. John had gone to bat for me.

He told me that our VP had wanted to eliminate my position then and to put me in an "individual contributor" role thus opening the new management slot in Harrisburg in January. John did not support that idea and asked our VP that this be postponed until the end of the quarter (and the year). He told him that I would probably make my numbers and that I would be a good candidate for the Harrisburg job in April. Our VP agreed to do that. That is why John had told me that the darkest hour was just before dawn and to just keep doing what I was doing.

I was flattered by all this, but it put me in an awkward position. John had protected my job and supported me for this promotion. However, I knew that Madelin and I would likely be no happier in Harrisburg than we were in Philadelphia. I thanked him for his support and told him that I would discuss it with Madelin and get back to him. It was another major crossroad in my career.

The next day I told John that we would still prefer to get back to the south if possible. He said he would work on it.

I didn't hear anything for a couple of weeks but finally, I received a call from HR telling me that there was an opening for a mainframe Branch Sales Manager in Richmond, Virginia. Apparently, they considered Richmond to be in the south, but I didn't, and I knew Madelin didn't either. As politely as I could, I told them that we were looking for something more like Atlanta or something even further south.

Richmond was the most southern point in Eastern Operations. Atlanta was in Southern Operations, which was headquartered in Houston. It finally occurred to me that Eastern Ops was not going to be much help in getting me back to Southern Ops. I decided to call my friend, Carl Straw, in Miami to see he could help. He was one of the only people I knew in Southern Ops, and he had helped me get the Philly job when the BC/7 division closed.

I called Carl and explained the situation. As always, Carl was very helpful. He told me that he thought there were sales management openings in Atlanta and Tallahassee and that he would confirm and get back to me. He soon called back to tell me that there were and that I needed to call Bill Grose, one of the Southern Ops managers in Atlanta. He said that he had spoken to Bill and that he was expecting my phone call.

I called Bill right away and had a great conversation. He told me a little about each of the openings and said that I could interview for either one, but not both at the same time. After talking it over with Madelin, we decided that Tallahassee sounded like the better option. Soon I was on a plane to Atlanta to interview with Bill and the Tallahassee Branch Manager, Ken Carter.

I arrived at the Sperry facility on the north side of Atlanta for the interviews and met with Bill first. He was very nice and engaging and spoke with a slow southern drawl that made me feel like I was back home. I left the meeting feeling that I had "checked the box." The real interview, however, would be with Ken over lunch.

Ken was a charming guy. He was funny, smart, and "all salesman." He laughed a lot and so did I. He told me about everyone in the Tallahassee office, especially the salesmen. He seemed to have a nickname for everyone. He called one of the guys "no-smoke" because he had asked Ken not to smoke in his car, and one of the guys "dirt" because he was older than everybody else. He made it a point to let me know that the sales rep in Jacksonville was a "Christian," something that Ken seemed to feel was some sort of notable characteristic.

He told me I would really like the systems manager, Darrell Wilson. When Ken started telling me who I would like, I knew I had the job. It is an implied close. Ken was assuming I would be coming to Tallahassee and wanted me to assume the same.

He was right about Darrell. I liked him and still do. It has been 38 years since that interview and Darrell is one of my best friends.

Ken was only in his mid-30s or so, but he was quite bald. He told me that he really liked Darrell because Darrell had even less hair than he did. He even told me about the local "watering hole" – the Sun Downer. He seemed proud that one of the cocktail waitresses there called them Shiney and Mop 'n Glow in honor of their baldness.

He described the three-member state sales team and how they worked together as a group, adding that I would really like Wayne Fountain, one of the team members. He was right about that and Wayne is still one of my best friends. He said that Wayne was new to sales and that he had worked in field engineering before. In Ken's opinion, Wayne was already the best salesman of the three and he predicted that Wayne would soon be leading the team.

He talked a lot about Spider Webb, the Sperry Univac lobbyist. To be honest, I had no idea why we would need a lobbyist. According to Ken, Spider was a great guy and could move mountains, but that he could "get you." I had no idea what that meant, so I just let it go.

Spider was in his 60s, old enough to be everyone else's dad. Spider and two members of the state team, Ray Pasternak and Jim Schrader have since passed away. They were all nice guys and I really liked them. May God rest their souls.

10

SOUTHBOUND

After Ken and I had long since finished our lunch, Ken asked me if had any more questions. I asked Ken if I had the job, to which he replied, "Hell, yeah! Let's get out of here." I was now the Branch Sales Manager for the North Florida Branch, responsible for the entire Sperry product line.

I returned home to Philly to begin arrangements for yet another move. This time we were going somewhere we wanted to go – Tallahassee, Florida.

A few days later, I walked into our house one afternoon and Madelin told me she had received a weird call from some "old-sounding dude" in Tallahassee named Spider. Spider had a thick southern drawl and spoke with an almost exaggerated politeness.

She said that he told her that he didn't work for Sperry but that he was some sort of consultant and that he had arranged for us to use a realtor in Tallahassee that was married to some important politician. I knew right away that this was headed in a bad direction. "What did you tell him?" I asked. "I told him that I didn't know who he was and that we preferred to choose our own realtor, but thanks anyway." It could have been way worse.

I called Ken to dig out and thankfully he thought it was funny. Spider had a good sense of humor about it too. Madelin and I always got along well with him. We ended up using his recommended realtor too.

We were both excited about moving to Tallahassee. Except for an 8-year stint in Jacksonville, I would spend the rest of my career there. Madelin and I were 29 when we moved to Tallahassee. That was 40 years ago, and we are still here.

We loved Tallahassee and still do. It is the polar opposite of Philadelphia. Madelin and I grew up in Mississippi, so living in Tallahassee was like being back home. Plus, we were now near both sets of Brian's grandparents, the original objective when we decided to leave the west. My parents lived in Fort Walton Beach, Florida, a three-hour drive and Madelin's parents lived in Meridian, Mississippi, a seven-hour drive. Both were close enough, but not too close.

The work environment was as different as the cities. Ken was engaged in everything, especially sales. He was always walking around the office talking with people or sitting in someone else's office. He always went to lunch with someone, usually a group. To the contrary, I never saw John Smith have lunch with anyone and, for sure, not with me. I spent more time with Ken during my first week than I did with John during my entire tenure in Philly.

On my first or second day, Ken told me that there was a company meeting in Daytona Beach for order entry people from all over the south. He said that we needed to go to make an appearance since we were the "host state."

We were also responsible for Orlando, Jacksonville, and points in between, including Daytona Beach. It sounded like a boondoggle to me, which it was, but he was my new boss, so we headed to Daytona Beach.

We rented a car and drove to Orlando so I could meet the people there, and the next day we drove to Daytona Beach for the meeting. Ken knew a lot of the order entry people, but we didn't do a thing except sit in their extremely boring meeting for a couple of hours and have lunch with them before heading to the airport. Ken's secretary had arranged for us to fly back from Daytona Beach.

Daytona Beach had a small airport and Tallahassee's was even smaller. Both airports were about the size of a Greyhound bus station and looked like one too. We had a direct flight to Tallahassee in a plane that held maybe five people. The pilot told us where to sit so that the load would be "balanced" and then turned around to tell us that there was a line of thunderstorms coming into Tallahassee. With his thick southern drawl, he said that he was pretty sure we could beat them. Ken looked at me laughing and said, "I bet you didn't know you signed up for this shit!" That was Ken.

I spent the next year trying to help my team, but I believe I learned more from them than they did from me. Selling to state governments, especially big states like Florida, is a completely different world. If anyone ever tells you that "selling is selling" beware of a huge caveat – selling is selling unless it is selling to the government. In that case, selling is something else.

Selling to state government is about the process, complex procurements, lobbying and political influence, and state contracts with terms and conditions that are, well, let's say "difficult." Decisions are sometimes more about political correctness than the best decision. In those days, large procurements were often not much more than proposal writing contests.

Sales axioms that generally apply in the commercial world do not necessarily apply to governmental sales. Even large companies that are well-experienced in government sales can make fatal mistakes by getting one or two people in the public sector sales management chain that do not understand these things.

When I started my job in Tallahassee, I thought I knew pretty much everything there was to know about selling. After all, I had been a successful sales rep in Phoenix and a successful sales manager in Las Vegas, Miami, and Philadelphia. What else could I possibly need to know? Well, as it turned out – a lot. While we did have some large commercial accounts, including Disney World and a few mainframe accounts in Jacksonville, most of our business was with the State of Florida.

In the commercial world, if the key decision-makers and influencers want to buy from you, they probably will. If your customers play golf and they play with you a lot, they are probably going to buy a lot of stuff from you. If you "shape the deal," you will probably win the deal. If the right c-level executive(s) want(s) to buy from you, everyone else will generally fall in line. Commercial clients can justify just about anything by declaring it integral to their "strategic direction."

Selling Information Technology

In state government, up can be down, and down can be up – especially in Florida's Invitation to Negotiate (ITN) world. The highest price might win, and the lowest price might lose. The best technical solution might lose, and the worst technical solution might win. The solution with the most risk to the state might win and the solution with the least risk might lose.

If you have worked on a deal for two years, and none of your competitors even knew about the deal until the bid came out, you could still lose. If you replied to a Request for Proposal (RFP) that you knew nothing about until you saw the RFP or ITN, you might win. If you have impeccable long-term client relationships, you might lose to a competitor who could not find his or her way from the airport to the Capitol.

Is it better to have great client relationships, great lobbyists, the best technical solution with the lowest price, and to have worked the deal more than anyone else? Of course. But does any of this mean you will win or lose? No.

I have lost plenty of state government deals that by all logic I should have won, but I have won more than my fair share of deals that I had no business winning. It works both ways.

I did not understand any of this when I started in Tallahassee.

I worked in the Tallahassee office for the next three years and had three different bosses: Ken Carter, Jay Taylor, and Carl Straw. Ken would be replaced by Jay within a year. A year or so after that, the Tallahassee Branch would be put back under Miami and Carl Straw would be my boss.

My first year in Tallahassee was quite a whirlwind. While I liked Ken personally and enjoyed socializing with him, to say he did not handle the pressure of being under quota very well would be an understatement. The further behind we fell, the more difficult it became to work with him.

He once told me that I should always be in the process of firing my worst performer on the theory that there was always somebody out there better than the worst guy. I just ignored that.

Ken was a salesman at heart and a very good one. He prided himself on client relationships and on structuring deals, and he was very good at both. But as the year went on, he became more and more of a micro-manager getting too involved in the daily activities of the sales reps. That was my job. I found myself in the middle of continuing conflicts between Ken and the sales team.

In the middle of the third quarter or so, Ken decided we would have mandatory Saturday morning sales meetings for the rest of the year or until we were back on our numbers. Worse yet, the meetings would be held in Tallahassee, Jacksonville, and Orlando. Ken volunteered to take the Tallahassee meetings and left it up to Darrell Wilson and me to handle Orlando and Jacksonville. As bad as it was for me, Darrell had it worse as he was not even in sales.

Everyone considered it as punishment and nothing more. It was double punishment for Darrell and me due to the time it took us to drive to Jacksonville and Orlando. For the first meeting or two, I tried to have an agenda of some things that could possibly be useful. That went nowhere and the meetings quickly turned into "gripe sessions."

Late one afternoon, Darrell and I ended up at Ken's house for some reason. Somehow, the discussion turned to a long-running feud between Spider Webb, our lobbyist, and the state sales team, especially Jim Schrader. By now, it was quite late, probably 9 p.m. or maybe later. Ken decided we would have a meeting right then and there to get all this sorted out.

This was clearly a bad idea and Darrell and I did all we could to stop it. Undeterred, Ken picked up his phone and started calling the salespeople –and Spider. In hearing Ken's end of the conversation, it seemed clear that Spider wanted no part of this. I think Ken may have woke him up.

Ken insisted and soon Spider was at Ken's house. Ken called Wayne Fountain and summoned him as well. No one answered the phone at Ray Pasternak's house. He only lived a few blocks from Ken, so Ken told me to go over to Ray's house and knock on the door – and the windows if I had to. Obviously, I did not do that. Someone called Jim Schrader and he reluctantly came to the "meeting."

The meeting was basically an argument over things that did not matter and that had happened years earlier. Nobody wanted to have this discussion, and everybody was unhappy about it. It was a non-event and we all soon went home.

News of the Saturday meetings and the late-night meeting at Ken's house soon made it up the management chain. Not long after that, Ken stopped coming into the office. He was still with the company, just not coming into the office. I don't recall any formal announcement and never knew exactly what happened. Maybe this is what Ken meant when he told me that Spider could "get you." I don't know, but I always wondered. I believe that Ken took a position with Sperry somewhere else, but I'm not sure about that either.

My zeal and determination to remain in sales management was wearing thin. I had done that for four years, in four cities and two companies. The only part of sales management that I really enjoyed was the selling part and there was too little of that.

I didn't enjoy being a backstop between my boss and the sales team and a referee between members of the sales team and other parts of the organization.

As a company, we spent too much time being internally focused, continually explaining to each other what we were going to sell and why we didn't sell whatever it was that we had previously said we were going to sell. Most of these time-wasting exercises fell on the sales managers. This phenomenon was not unique to Sperry and it carries on today with most IT sales organizations. Generally, the further up the management chain you go, the worse this phenomenon becomes – but I digress.

I was tired of it and needed a break from management. At the end of the year, I asked to be reassigned from my sales management position to a sales executive position.

In my new position, I was responsible for selling data entry computers to the Florida government and the entire product line to commercial accounts in the Tallahassee area and Jacksonville. Wayne Fountain was promoted to my previous position, which was great as he had become a good friend. Jay Taylor was named the new Branch Manager and I liked him as well. He had been in Tallahassee before and came back to take this assignment.

This was a much-needed change for me, and the coming year was much better. I sold a lot of 1900/10 computer systems to the State of Florida. One agency, the Department of Highway Safety and Motor Vehicles bought five systems as the result of a single procurement. I sold systems in Valdosta and Jacksonville as well. I was much happier with less stress and more money.

Selling Information Technology

While it was a great year for me, it was almost a super great year. Selling a "new account" 1100 mainframe was the crown jewel for the Sperry Univac sales organization.

Davis Water and Waste Industries was a mid-sized distributor of industrial irrigation systems, products, and supplies in Thomasville, Georgia, about 30 miles north of Tallahassee. They had distribution locations across much of the Southeast.

I contacted them by making cold calls from a Dunn and Bradstreet listing and reached their CFO, Stan Dubyew. He was polite and quite willing to talk with me. He was eager to tell me that they had an aging IBM system that they had outgrown and would soon be replacing. They would be modernizing their order entry, billing, and inventory system. He agreed to meet with me to learn more about Sperry Univac and what we had to offer.

Over the coming months, I had several meetings with Stan and other members of the Davis management team. Most of these meetings were in Thomasville, but Stan and other Davis executives came to Tallahassee several times for formal presentations and demonstrations. Each step along the way increased their interest in our 1100 system and MAPPER, a proprietary software system that ran on the 1100 series of mainframes.

MAPPER was a software system that was designed to make it easier for non-programmers to generate custom reports. MAPPER was an acronym derived from Maintain, Prepare and Produce Executive Reports. This was before PC-based spreadsheets like LOTUS 1-2-3 and Microsoft Excel but it was somewhat of the same idea. It was very popular with 1100 customers. Most 1100 demos included MAPPER demonstrations and Davis was no exception.

MAPPER was most appreciated by advanced users that were analytically inclined. Stan Dubyew and a few others there fell into that category. Stan particularly liked MAPPER. We completed several demos using dummy files of Davis's, part numbers and descriptions. All of these demos went well. It was MAPPER that sparked their interest in Sperry Univac.

However, as much as they liked MAPPER, they would still require an advanced order entry, billing, and inventory software solution that would run on the 1100. We didn't have one and were struggling to come up with something. It was, essentially, the only thing that stood between us and an 1100 sale to Davis.

Finally, we found a potential solution, an application called Wizard, that had originally been developed for Data General systems but partially converted to run on Sperry System 80 computers, a smaller cousin to the 1100. The idea was to prove the functionality to Davis on the System 80 and then convert it to the 1100.

We presented this idea to Davis and gave them some basic documentation on Wizard. They approved the plan, and we immediately began shoring up the Wizard application on the System 80 so that we could begin software demonstrations. The first couple of presentations went well enough that Davis could see the potential. However, there were still significant problems. They knew it was a work in progress, so they were willing to accept the "imperfections" at that point.

Somewhere during this process, Stan Dubyew called me and said he wanted to meet with me to discuss a few things, assuring me that it was all good. I met with him soon afterward and he told me they were ready to look for a new

Director of Data Processing. To my almost disbelief, he told me that they wanted to hire someone with an 1100 background and asked me if I could help identify some qualified candidates. The news does not get any better for a computer salesman.

Stan did his best to curb my excitement by telling me that there were no promises and that there were still a lot of bridges to be crossed. Despite his best efforts, it was clear that they were planning on leaving the IBM family and buying an 1100. I left the meeting extremely excited knowing that they were headed toward an 1100 decision – and I knew the perfect candidate.

When I got back to the office, my first call was to my good friend, Darrell Wilson, in Houston, Texas. Darrell had been promoted to a Sperry Univac Southern Ops management position in Houston after having served for several years as the Systems Manager in Tallahassee. He was experienced with the 1100 product line and was an excellent manager. I knew that Davis would want to hire him, but I didn't know if he and his wife Diana would be interested in coming back to the Tallahassee area and in leaving Sperry, but I had high hopes.

I talked with Darrell and to my great relief, he was interested in finding out more about the job. With Darrell's ok, I floated his name and background to Stan Dubyew at Davis. As I knew he would, Stan thought Darrell was the ideal candidate. Within a day or two, Darrell was in discussions with Davis Industries, and he soon became the new Director of Data Processing for Davis.

Darrell and Diana bought a beautiful new home in Thomasville. Davis had insisted that they live in Thomasville and not Tallahassee some 30 miles to the south. Apparently, Davis had had previous bad experiences with people living in Tallahassee and commuting to Thomasville. They wanted to know that Darrell was committed not just to Davis, but to Thomasville as well. Except for the struggle with Wizard, this could not be going any better.

There was always tremendous self-imposed pressure to win deals. But this time, it wasn't just about me and my company. It was also about my good friend, Darrell.

While Darrell could be successful there under almost any circumstance, he was there because of his Sperry background, because of Davis's intention to transition to a new modernized business environment based on Sperry Univac technology, and because I put Darrell and Davis together. "Landing the plane" was equally important to both of us.

While all of this was going on, there were lots of other things going on in my life as well. For one, our daughter, Jamie, was born on September 23, 1983.

I was supposed to be in Jacksonville on the 23rd for an early morning meeting. Jamie wasn't due for another week so Madelin assured me it would be ok for me to go over for the meeting.

I drove over the night before to stay at the Sea Turtle Inn in Jacksonville Beach and, sure enough, just as I fell asleep around midnight, the phone rang. It was Madelin telling me that she and her mom, Bernice, were heading to the hospital.

I got up, took a shower to wake up, and started the drive back to Tallahassee. I got to her room at Tallahassee Memorial around 4 a.m. and found her in full labor up on all fours in the hospital bed with wires hooked up and sort of mooing. The first thing she said was, "Hi, don't touch me," which was about the way it went when Brian was born six years earlier.

Jamie was born at 10:41 that morning. She was a beautiful, healthy baby girl and Madelin did just great. I saw Jamie come into the world and take her first breath. After a couple of hours, Madelin gave me the ok to take Bernice home with a list of things to bring back for her. We stopped by my office so I could hand out "It's a girl!" cigars and order flowers for Madelin and Jamie.

Within a few days, we brought Jamie home. We were one big happy family with two kids and a cat named Kitty. Just as it was with Jamie's brother, we would move a few months after she was born. September 23, 1983, was one of the two best days of my career.

I was soon back at work, trying to close the Davis deal, managing a few existing accounts in Jacksonville, and working a growing list of prospects. It seemed that I was either in Thomasville or Jacksonville just about all the time. I was spending so much time in Jacksonville that Sperry rented an apartment for me, and I was there more than I was at home.

My company had been wanting me to transfer to Jacksonville for a couple of years, but we were tired of moving. Finally, even though we both loved Tallahassee, we decided that it would be best for our family and my career to take the transfer to Jacksonville. Jamie was just 6 months old, and Brian was 6 years old. We would be in Jacksonville for the next 8 years.

For the next several months, I spent most of my time working on the Davis deal. Darrell was now on board as the Director of Data Processing. The Davis management team liked him, and Darrell liked his new job, although I think he was as stressed as I was over the pending 1100 deal.

Before Darrell started, Davis had agreed that they would purchase the 1100 system if we met certain conditions regarding the Wizard application. In a nutshell, we had to prove that it would work. They gave us a letter of intent but nothing that we could book. We knew exactly what we had to do.

We were now completely focused on making Wizard work, at least well enough to have a successful demonstration. We had one of our best programmer/analysts working on it full-time. Within a month or so, we were ready to have a trial demo for Darrell – and only Darrell. Once he gave us the ok, we would then schedule the "real" demo for the rest of the Davis management team.

Darrell came in for the demo and it was a disaster. The simplest of things did not work. Quantity on hand, 10. Sell 6. New quantity on hand, 5. That sort of thing.

He was not one to mince words, especially with people he knew well. He made it clear that a repeat of a demo like this for the management team would spell disaster for the 1100 sale, for his credibility, and possibly for his future at Davis.

There was no doubt in anyone's mind as to the critical importance of Wizard. Still, it was becoming more and more clear that Wizard may be unsalvageable. It was not well-documented and some of the programs had never been fully completed. Unstable was an understatement, it was as flakey as a box of Cornflakes. We needed an entire team of developers and testers, but that was not what we had – we had one person.

Within a few weeks, we had corrected all the known

problems and felt cautiously optimistic about a successful demo.

We scheduled the "real demo" to be held in Atlanta. I flew there with the Davis management team on their private plane. This was a big event and the most important demo of my career at the time.

To say the least, it was not what he had hoped for. We managed to complete the demo but hit several snags along the way. The Davis people were polite but not enthusiastic and not yet ready to ink the deal. We were hoping for "Great, let's get the contract completed and get to work." Instead, it was more like "We'll get together and discuss where we are and get back to you" – not the best of buying signs.

It was now near the end of our business year. Davis at least had agreed to review the final contract and put it on their executive meeting agenda for discussion for the last week of the month. We were not enthusiastic about a positive outcome but at least there was a chance. I traveled to Thomasville on the day of the meeting hoping against hope to leave with the signed contract.

When I arrived, the meeting was still in progress and, to my surprise, the meeting room was adjacent to the lobby. I could hear some of what they were saying, enough to know that it was not going well. When the meeting was over, I met with Darrell and maybe Stan Dubyew. I was told they were not yet comfortable with signing a contract due to their lack of confidence in Wizard. We still had more work to do.

I left the meeting disappointed but not surprised. I drove to the Tallahassee office to advise my managers, Wayne Fountain, and Carl Straw, that we would not be closing the deal before our year-end. Like me, they were disappointed, but we still had a chance for the coming year, yet not a particularly good one.

Predictably, we never recovered from the Wizard fiasco and never made the sale to Davis. As the saying goes, it was ours to lose, and lose we did.

As I had learned during my first year in sales, no one wants a computer, everyone wants a solution. What Davis really wanted was a new order entry, billing, and inventory system, not a new mainframe computer. If we would have had an acceptable software solution, we would have easily closed that deal.

Darrell stayed with Davis for a while but eventually left to take a position in Tallahassee with the State of Florida Information Resource Commission. We would work together again in the coming years at DEC and Unisys and play hundreds of rounds of golf together.

I beat him one time and have the scorecard to prove it. It's been over 35 years since all this happened at Davis and we remain very good friends.

11

RETURN OF THE PHOENIX CONNECTION

A few months later, Madelin and I decided to take the kids on vacation to Scottsdale, Arizona. Disney World and a thousand other places would have been a better idea, but Scottsdale seemed like a good idea at the time. It was more nostalgia for Madelin and me than anything else.

Unfortunately, it was in the middle of summer and it was hotter than hell. Selective memory had set in. We could only remember the things that we loved about Phoenix as we filtered out all memory of the brutal desert heat.

We took the kids to the Phoenix zoo one morning. It was extremely hot, which is maybe why we had the entire zoo to ourselves. Staying in the indoor exhibits helped but we still ended up on the verge of heat strokes. Another day, I took Brian to play miniature golf at the Metrocenter Mall and the same thing happened.

We finally gave it up and stayed in the hotel or did something indoors during the days. Even the pool at the hotel was too hot to swim in during the day. Madelin griped at me for years about this vacation, insisting it was all my idea.

While we were there, I called Hans Ager, my friend, and business colleague that had founded Applied Data Systems (ADS) when we lived in Phoenix. I wanted to say hello and to see how he and his company were doing.

He was as friendly as always and was glad to hear from me. He had moved his company into a new facility in North Scottsdale and invited me to come out for a tour.

Hans was the best salesman I have ever known. If you didn't want to buy what he was selling, you should stay away.

I made the short drive out to ADS the next day. He gave me the grand tour of their impressive facility, introduced me to the executive team, and then took me to his office to catch up. It was as if he had planned his entire day around our meeting.

I told him all about Brian and Jamie and the four corporate moves we had made since we had last seen each other. He was as engaging as always and asked me a lot of questions in his inquisitive but friendly way. He had an uncanny ability to make everyone feel like they were especially important to him – a trait of a good salesman.

He told me about everything that had happened with ADS over the last six years and how ADS was now totally focused on heavy equipment dealers. They had switched from Burroughs to Texas Instruments (TI) for their main computer platform. He excitedly told me how ADS had signed national agreements with John Deere and Caterpillar and about their expansion plans that included the creation of an "elite" nationwide salesforce.

He soon told me in his confident style that I should join them. He explained that they would soon be opening a southeast regional office in Atlanta and that he would like me to be the regional manager.

I was flattered but told him that I was happy at Sperry and that moving again was completely out of the question. He quickly replied that I needed to get out of Sperry and that he had no problem with Jacksonville becoming the regional office, instead of Atlanta.

We talked about it for a while longer, and the more we talked the better it sounded. I told him that I would discuss it with Madelin and get back to him.

As soon as I got back to the hotel, I told Madelin all about the meeting and about the possible job. As I knew she would, she told me that she was fine with whatever I wanted to do, as long as we didn't have to move.

Despite all the heat and the near heat strokes, we had a good time on the rest of our vacation. We took the kids to Rawhide, a simulated old-west town north of Scottsdale, to Carefree, and to the top of South Mountain. We also drove by the house where we lived when was Brian was born. I don't think the kids were particularly impressed with any of this, but they enjoyed the swimming pool at the hotel and the plane ride.

Hans and I continued to talk by phone, and within a few weeks, I had a formal offer to join ADS as the Regional Sales Director for the southeast.

ADS was a small company with probably less than 100 people. I would not be working for a Fortune 500 company anymore, and it was not without risk. ADS was a niche company with one product for one industry, but they were very good at what they did.

It was possible that ADS could become the next ADP or Reynolds and Reynolds, both giants in the dealership IT space at the time. If this were to happen, I could be rich and retire by 40, but there were plenty of things that could go wrong too. I was only 34 and there would be plenty of options and plenty of time if it did not work out.

The biggest downside was leaving Sperry, a company that I really liked and that had been good to me. I had lots of friends there too. The worst part was telling my boss and good friend, Wayne Fountain, and his boss, Carl Straw, also a good friend. I felt particularly bad about telling Carl after all he had done for me helping me get the Philadelphia and Tallahassee jobs.

They were both nice about it, as I knew they would be, but both did their best to talk me out of it. Once that didn't work, Carl asked me if I would fly to Atlanta to talk to his boss, Al Green, before making the final, final decision, and I agreed to do that.

I flew up one morning and met Al at an airport coffee shop. He said all the same things that Wayne and Carl had already said but added something to the mix that was quite out of the ordinary. He offered to pay my commissions in advance for the Davis deal if I would stay. Unfortunately, this deal was hanging on by the thinnest of threads, and short of a miracle, it was not going to happen. Something that I guess Al did not know.

Still, it was a generous offer. If I had stayed and accepted their offer, I would have probably ended up being overpaid by Sperry for around $25,000, a lot of money at the time. I stayed with my decision and resigned.

For the previous ten years, I had worked for large corporations, in large offices, with lots of people. ADS would be much different.

My first ADS task was to lease an office. Hans wanted an impressive office in a high-end business park. We rented an office suite that offered secretarial services in the Southpoint Office Park in the Baymeadows section of town. It was perfect. They screened all my phone calls, took messages when I was out, and did all my typing for proposals and such things.

I spent the first few months getting settled into the office, traveling to Scottsdale for product training, learning to demo the software, and doing some basic prospecting. While ADS would occasionally fly someone in for a big demo, most of the time I had to do that myself.

The Texas Instruments computer we used was not exactly portable. It was about the size of a small file cabinet and was very heavy. Knowing that I would have to learn to demo the software at night and on the weekends, I set the computer up at our house. That turned out to be a good plan. After a few weeks, I moved the computer to the office.

My first demo was with a tractor and implement company in Forsyth, Georgia. I loaded the computer into my Jeep Cherokee and headed to Forsyth some four hours northwest of Jacksonville.

I was quite nervous and unsure about it. Although I had been selling computers for ten years, doing my own demo was not typical for me. Not only was it my first ADS demo and but it was also my first time to meet this customer other than talking with them on the phone. Fortunately, they were very nice and rolled out the southern hospitality.

There was a room full of people including their outside CPA, the dealership managers, and the owner of the dealership, all of whom wanted to see every single thing the software could do.

The ADS software was very good, and I was able to show them almost everything they wanted to see and to answer most of their questions. To be safe, an ADS software analyst was on standby in Scottsdale for questions that I could not answer.

I left feeling good about the demo and confident in my new ability to demonstrate the software. The owner of the dealership told me he appreciated my answering his questions rather than saying "gotcha covered" to everything, which was apparently what my primary competitor had done. It took a couple of more trips to Forsyth, but this ended up being my first ADS sale.

From there on, every demo and every sale were pretty much the same. Every dealership would ask the same questions. They always had the same objections and always liked the same things. The competitors were always the same, and there were only a few of them. I knew their weaknesses and always made sure that my prospects knew them as well. Even though ADS was a relatively young and small company, we were generally considered to be the leading provider of dealership software. I rarely needed to explain who we were – but always did. While I had some prospects that "decided not to decide," I rarely lost a deal to a competitor. ADS was a big fish in a small pond.

It was fun being the king of the hill and winning most of the time, but the travel got old. I was on the road all the time. I covered the Carolinas, Georgia, and Florida. Even though I was hired to be the sales director for the southern region, I was essentially a sales rep covering four states. I had two young children that did not like me being away from home, and Madelin was not crazy about it either.

This would get better, theoretically, once I had a sales team on board.

I soon made my first ADS hire. He was a few years younger than I, around 30, but had several years of experience selling to dealerships. I hired him to cover the Florida territory.

It did not take long to see that he had little interest in learning the ADS software. I suggested to him numerous times, more than subtly, that he take the demo system home to learn the software at night and on weekends as I had done. He just would not do it.

For the most part, he was not much more than a traveling companion and someone to help me load and unload the demo computer. He never sold a thing and was gone after a few months or so. I was back at square one.

It was during this time period that I had my vasectomy. Being an RN, Madelin had made a compelling argument that it would be a simple procedure. Not that I didn't trust her, but I called every guy I knew who was a vasectomy veteran.

Every one of them told me the same thing. It was no big deal, a relatively painless procedure, and that within a couple of weeks everything would be back to normal. Either they all lied to me or my doctor was a quack. I think it was the latter.

After the procedure and once I felt good enough to even talk on the phone, which was several days, I called every one of them. They all said the same thing again, adding, "You mean they didn't sedate you at all?" I could write a whole chapter on this, but you get the idea.

Anyway, one of the post-op instructions was to avoid lifting anything heavy for two weeks.

Two weeks after the procedure, ok, maybe a few days short

of two weeks, I had a demo scheduled in Fort Lauderdale. Madelin felt that it was too early for me to drive that far and to be lifting the computer, but I went anyway.

I had my neighbor load my computer into the back of my Cherokee and I headed south for the demo. I had a new light blue suit and wore it for the first time for this demo.

When I arrived, I had someone come out to help me unload the computer. It was not a one-man job, so together we lifted it from the back of the Cherokee onto the hand truck. Within half an hour or so, I was set up and ready for the demo.

The demo was in their conference room, which was full of people, mostly women. I sat in the middle chair at the conference table so that everyone could gather around and behind me to see the demo – no big screens then.

It was a long demo, at least two hours, but it went well. After the demo, we took a break before regrouping for questions.

I got up and walked into the restroom where I immediately realized that I was bleeding like crazy. My new light blue suit was soaked with blood from my crotch halfway down to my knees. But it got worse. I had also ripped out the seat of my pants from the back belt loop to the zipper. I guess both rips occurred unloading the computer.

My first thought was to just stay in the restroom until they all went home – but that wouldn't work. I did my best to stop the bleeding, got up my courage, and opened the door. I was torn between tying my suit coat in front of me to cover up the bloodstain and tying it behind me to cover up the huge rip in the seat of my pants.

They must have noticed all of this when I got up to walk to the restroom, but they didn't say anything. I apologized and explained to them that I had recently had minor surgery and that something had happened – and that I also had ripped my pants.

If it had been men, it would have been a big joke and everyone, including me, would have laughed it off. But it was mostly women. They were all sympathetic and even offered to drive me to the hotel or the hospital. The men loaded the computer for me, and I drove to my hotel.

I called my doctor first. In addition to being a quack, he was an asshole. He told me, in an agitated fashion, that he could not do anything for me in Fort Lauderdale and to go to the emergency room if it was bad enough.

I then called Madelin. She was somewhere between concerned and pissed off. She told me to order room service, stay in bed with ice between my legs, and drive home the next day if I was able. Like the doctor, she also told me to go to the emergency room if the bleeding did not stop. I followed her directions and drove home the next day.

Within a month or so, I closed that sale. If nothing else, I had won their sympathy votes.

For about a year, everything went fine. ADS hired Greg Shortell, a Motorola executive, as the corporate VP of Sales. He was a sharp guy with a lot of good ideas. He had an impressive background, so impressive, in fact, that I had to wonder why he had taken the job. He just seemed overqualified to me, but Hans probably promised him the moon.

He soon traveled to Jacksonville to spend a few days with me, no doubt to form his own opinion of me. I liked him and felt like he had a lot of good ideas for the company. I think he liked me too.

He laid out his plan and vision and seemed genuinely interested in my opinion on things, and I have always had an opinion. What I liked most about him was his commitment to building what he and Hans called "an elite salesforce" and his assurances that I would play a key role in the future of the company.

I continued to travel a lot as it was a necessary evil. Remote demos that are commonplace today were not practical then. Occasionally, I would take Madelin and the kids with me just so we could all be together. Brian and Jamie were young and happy to go anywhere if I promised that the hotel would have a big pool.

I traveled to Columbia, South Carolina, quite often and I took Madelin and the kids with me on one of these trips. We had so much stuff to take that I bought a luggage carrier for the top of our Jeep Cherokee.

When we arrived, we unloaded the carrier, took everything into the hotel, and locked the carrier back up.

As I walked up to the Jeep the next morning, I notice that the top half of the carrier was gone. Someone had broken the lock to steal whatever we had in there. I was quite upset that someone had done that, but at the same time felt a great sense of satisfaction knowing that they got nothing for their troubles except the top half of the carrier which they stole for some weird reason.

At least once a day, we ate at Lizard's Thicket, a local restaurant that was great for families with little kids. Madelin and I also enjoyed reading the bumper stickers and tee shirts for the South Carolina Gamecocks. Our favorite was "You Can't Lick Our Cocks." We still laugh about that sometimes. The kids were too young to care and too young to get it. As a side note, while catchy, the bumper sticker was not factually correct. Most teams did lick their Cocks and most still do.

Selling Information Technology

After a few days, we drove home with a tarp tied over the open luggage carrier which seemed like a good idea at the time. However, it was summer and predictably it rained on the way home. In essence, we had a kiddie pool on top of the car with floating luggage.

When we got home, I drilled holes in the carrier to drain the water. Despite all of this, it was a fun and memorable trip. And my demos all went well too.

For the next two years, I drove thousands of miles and sold a lot of systems for ADS. I was in my own little bubble and became somewhat oblivious to the sweeping changes that were happening in the computer industry. One of these was the monumental change that personal computers were bringing to the computer industry and to corporate America.

The first time I ever encountered a business that was seriously considering personal computers to solve a real business problem was when I was with ADS.

I had a prospect in North Carolina that I had communicated with only by phone and mail. I was pretty sure that they were going to buy something and if they did, it would probably be from me. I was confident and for good reason – nobody had better software than we did.

I called them one day to let them know I would be there soon in hopes of scheduling a demo.

The owner of the dealership told me they had decided to implement a PC LAN (Local Area Network) and to have someone custom develop the software. I knew nothing about PC LANs other than a vague idea of PCs hooked together. PCs were still considered to be mainly home computers and, at best, some sort of nerdy business experiment.

As far as I was concerned, his decision was just plain stupid.

I tried to convince him that even if the PC LAN idea proved to be workable, the chance of someone custom developing acceptable software was somewhere between slim and none. He would not listen though. My guess is that they never got off the ground, not necessarily because of the PC LAN idea, but because of the complexity of developing the software.

But indeed, there was an IT revolution in process around the advent of the IBM PC and the explosion of PC-based LANs.

Novell introduced corporate-level networking software in the early '80s shortly after IBM released their first PC. IBM released the IBM PC AT around 1984, and client/server computing became the new buzzword and a real architecture for software design. Up until this point, computing was done by a central computer with very little intelligence in the workstations, which was the way the ADS systems worked.

PCs and local area networks would have a profound effect on the computer industry that would unfold for the remainder of the 1980s and beyond.

This was the beginning, at least in part, of what became the move to "open systems." Other computer companies began to release their own PCs. Everybody jumped on the IBM PC-compatible bandwagon.

Arguably, even IBM PCs were not "IBM" PCs. They were PCs assembled by IBM using Intel and other hardware components and a Microsoft operating system. Traditional computer companies like Sperry Univac and Burroughs and the new startups like Compaq and Dell released PCs doing the exact same thing.

The move to PCs was a boon for Intel, Microsoft, Novell, and PC software companies, but a disaster in the making for most of the traditional computer companies including IBM. While the computer industry was falling all over itself to be part of the "me too" PC revolution, there was at least one notable exception – Digital Equipment Corporation (DEC). DEC did not follow the PC parade – at least not with any conviction or enthusiasm.

Most of the traditional computer companies did not know what to do with PCs. The challenge was not in manufacturing but in sales and distribution.

Some of the new startups, like Compaq and Dell, figured out how to do it. Further, they started with a clean slate not having to worry about cannibalizing other more profitable product lines. They created a new and efficient way to sell.

IBM and others had created their own worst nightmares. PCs were eating into the sales of more profitable and proprietary computer systems. Additionally, IBM and the other traditional computer companies could not figure out how to distinguish their PCs from anyone else's. They did not know how to sell in high volumes or how to make a profit.

PCs soon became a commodity and selling PCs became a race to the bottom for most computer companies. After spending years trying to get into the PC business, these companies would spend years trying to stop the bleeding and trying to get out of the PC business. PCs were here to stay, but not with IBM, Sperry Univac, and other such companies.

Except for the one deal in North Carolina, none of this was affecting me, but I'm sure that Hans and others at ADS corporate could see the handwriting on the wall. ADS would have to move the software to a client/server environment using Intel and Microsoft technology. Eventually, the market for systems anchored to old minicomputer technology using dumb terminals would begin to erode.

After a while, it became obvious that something was wrong at ADS.

At first, we all noticed that the payment of expense reports was getting slower and slower. Soon, my office landlord began calling me about our rent being late. Rumors started of ADS being acquired. Still, I felt that if there was something I needed to know, Hans would give me a heads up. I was wrong about that.

It was not long before we were all informed that payroll checks would be delayed for a week with ADS offering up some lame excuse. I called Hans and he tried to put a good spin on an obviously bad situation telling me that ADS was within days of signing a major capital venture deal that would allow us to become a much larger firm. He assured me that I would be very happy with the coming news and that I would play an important role in the future of the company.

12

DIGITAL HAS IT NOW!

Several of my previous Sperry Univac colleagues were working for Digital Equipment Corporation (DEC or Digital). They called me quite often, always raving about DEC and telling me that I should join them. I called one of them and within few days I received a call from Bill Hughes, one of DEC's Florida managers.

DEC was in a hiring frenzy and was the darling of the IT industry. The Digital logo and glowing press were splashed all over newspapers, business magazines, and trade publications. They had several catchy (and accurate) taglines like "Digital Has It Now!"

As I would soon learn, all of this was for good reason. DEC had amazing and elegant technology, like nothing I had ever seen. On a scale of 1 to 10, if midrange computer technology from Sperry Univac, Burroughs, and IBM were 5s, DEC was a 9 or 10 – in my view.

The "Digital Has It Now" tagline was aimed at IBM, and everybody knew it. IBM had fallen behind in technology, especially in midrange computers. Their various divisions had created numerous overlapping systems that were not compatible with each other. This was especially true in midrange computers where DEC excelled.

It seemed that IBM was living off its reputation, always selling how great the next system would be and always avoiding discussions about their current offerings, the products that customers could actually buy.

It was time to leave and move on. I resigned from ADS in 1987 to join DEC at the peak of their amazing rise in the computer industry.

Revenue growth was outpacing most of the industry and profits were soaring. DEC's senior management would boldly predict at corporate meetings the exact month and year that DEC would surpass IBM as the world's largest computer company.

DEC was starting a new nationwide sales organization chartered to pursue IBM accounts in Fortune 500 companies. They described this as "breaking into the glass house."

Despite all its success, DEC still had a small share of corporate systems that ran the core business applications of large corporations. DEC's success was mainly in departmental, engineering, and technical systems. DEC believed it was time to take on IBM in their own backyard – the datacenters of the world's largest corporations. It was the new frontier.

To do so, DEC created a new sales organization they called the Information Systems Sales Team (ISST). This sales organization was built mainly from the outside, with people that had been successful selling large systems to large accounts.

I joined DEC as the ISST Sales Executive in Jacksonville, Florida, and was assigned Barnett Banks of Florida, Blue

Cross/Blue Shield, and CSX Corporation. All three of these were IBM bastions and none of them knew, or cared, who DEC was. My boss, Bill Hughes, was a DEC veteran based in Orlando. Bill established and ran the Florida ISST organization. He was a great guy and manager, and I really enjoyed working for him.

One of the things that appealed to me about DEC was a dual-option career track for sales executives. In most computer companies, to move up in the organization, you had to accept sales management positions with each step moving you further away from client-facing sales. DEC had a sales organization structure whereby you could stay in frontline sales and still move up in the organization.

As a Sales Exec, I could someday become a Sales Exec III managing the Ford Motor account in Detroit worth hundreds of millions per year in revenue, as an example. This had great appeal to me.

DEC's internal operations were more advanced than anything I had experienced before. On my first day, I was given a computer terminal and an email account on DEC's ALL-IN-1 office system. It was the first time I ever used email.

I received some basic training including how to look up any employee in the company on ELF (the employee locator facility) and how to send emails. As strange as it seems, few organizations had email systems at that time. I was shown how to access various administrative systems including online documentation and sales materials. It was also the first time in my career that brochures and manuals weren't kept on shelves in an office library.

They also gave me a brief explanation of DECnet, explaining that there were 60,000 nodes on DEC's worldwide internal network. To say the least, I was extremely impressed and went home after my first day excited about my new job.

As a side note on ELF, one day my DEC buddies and I keyed in "find Jack Shit" into ELF. I guess we were bored. ELF came back and said, "Could not find Jack Shit, did find Jack Shields." Jack Shields was one of the top DEC executives. We found that to be hilarious. And by the way, that was quite an impressive search technology for the day.

One of my first tasks was to attend DECworld in Boston in the fall of 1987. This was a massive and glitzy trade show of DEC's technology. At the time, it was probably the largest trade show of its kind.

DEC spared no expense. They even chartered two cruise ships, one being Queen Elizabeth 2, and anchored them in Boston Harbor to serve as extra hotels. The event was attended by more than 50,000 executives and IT professionals from around the world.

I always hated going to trade shows and conferences, and I would not have gone to this one had my boss not told me I had to. But it was the perfect introduction to DEC. I had never seen anything like it. Everything was first-class and there was excitement there among everyone.

The entire complex was networked into DEC's internal worldwide network. DEC technology was everywhere, from VAX-based graphics workstations to large VAXcluster computer systems. Also, when attendees registered, they were given a DEC email account so that they could stay in contact with their colleagues while they were there via DEC terminals placed throughout the venue – a brilliant marketing ploy to help everyone see the value of email. No one had cell phones, and few had email back home. DEC had the best office automation software in the industry – ALL-IN-1.

"They are renting a big boat to get mass media coverage - they want executives who don't think about technology to know who they are," said Susie Peterson, an analyst with the First Boston Corporation. "D.E.C. thinks it is I.B.M.'s equal and wants Fortune 500 buyers to agree."

Over the coming years, DEC put the DECworld concept on the road, trucking networking, and computer equipment to major cities for mini-DECworlds. These were impressive as well. Additionally, DEC opened permanent upscale demonstration facilities called Application Centers for Technology (ACTs) in major cities with the nearest one to me being in Atlanta. DEC had never been known as a marketing company but in the 1980s no one out-marketed DEC.

I observed that DEC was also more culturally and socially "advanced" than Burroughs and Sperry Univac. DEC was a noticeably diverse company. While the sales organization was still "youngish," there were many more women, African Americans, and other minorities in sales and management positions than I had seen in the other companies.

One of the interesting and unique cultural things I noticed was the large number of men with full beards – I'm talking ZZ Top beards in some cases. That maybe came from DEC's roots as an engineering-based company. But for whatever reason, there were lots of men with beards, something you seldom saw at Burroughs and Sperry Univac.

DEC was also very inclusive of spouses, always including and recognizing them at company achievement and celebration events. It was a nice change, but to be fair, this was changing everywhere in the American business culture in the '80s, but DEC seemed to be ahead of the curve.

I began to do as much research as I could on CSX, Barnett, and Blue Cross. It was not easy as there was no Google. I didn't know a thing about these huge companies and neither did anyone else at DEC.

I was able to gather some annual reports, brochures, and other publicly available information (all hardcopy) and that was about it. I made lots of phone calls to their public relations departments asking for key names and any information I could get. I soon started making phone calls to the firms, mostly looking for names of high-level potential contacts.

At the same time, I was preparing for DEC's two-week computer systems sales training class in Boston. DEC was a highly technical company, and to them, the product was everything. They shipped me a box of mundane technical manuals to study for the course. It was obvious to me that this information would be irrelevant to my job, but I studied the material anyway.

There were manuals on printers, networking gear, processors, storage, software, operating systems – everything. They believed that we should know "off the top of our head" the most appropriate printer for a department that required a matrix printer capable of printing on 3-part paper. There was not a Fortune 500 CIO or C-level executive anywhere in the world that cared about such things, but you could not convince DEC of that.

The class was two weeks of pure hell. The class material and lectures were more boring than the manuals they had sent me. We had a class administrator that was basically the person that called the roll and told us when we could go to lunch. The instructors were mostly engineers from nearby DEC plants who routinely rolled in gear with the side panels off to show us what a "BI-bus," or something, looked like.

Most of the people in the class were big-picture people like me, so there was plenty of eye-rolling going on amongst us. And, by the way, the "BI-bus" looked like a piece of metal with a bunch of wires attached to it.

Each day started with a quiz on the previous night's study material. They were serious too. If you didn't maintain a certain score, you would be sent home, and several were.

One of my classmates was Mike McSmith, a nice guy from Tampa who was also part of the Florida ISST team. We became good friends.

Mike was a fun guy. He was tall, blonde, had a good suntan, and looked like a "Floridian." He rented a convertible to use while at the class. I rode with him back and forth to class every day, always with the top down. He could have been a double for Don Johnson in Miami Vice.

While we were there, several of us went to two Red Sox games. The first one was a Sunday afternoon game. Mike wanted to sit in the outfield bleachers so we could watch the drunks get thrown out. I thought he was kidding, but he wasn't. There were lots of obnoxious drunks, indeed, and quite a few of them got thrown out for being drunk and rowdy.

For the second game, we decided to try for infield seats. We went to the box office just before the game to get our tickets. I am sure the ticket dude was messing with us, but I did not know it then. We asked if there were any infield seats and with his grumpy demeanor and Boston accent he said, "Yeah, I got two seats behind 3rd base, but one has a limited view." He said that it had a beam between the seat and the field that would restrict the view. That sounded doable so we said ok.

The beam was directly in front of my seat to the extent that I had my knees on each side of the beam. It restricted my view alright, all I could see was the beam. We laughed our asses off, drank our beer, ate our hotdogs, and then moved to two open seats for the rest of the game.

13

TUNNEL OF DARKNESS

But it wasn't all fun and games – we still had work to do. As was typical for such training sessions, we were separated into groups near the end of the class for our final presentations to a panel of instructors.

My last 18 hours there changed my life for the worse for the next five years, harmed my health, and threatened my career. A darkness fell over me that remained for years. But the light eventually returned.

On our last night there, I met with my assigned teammates to work on our final presentation. It was a routine assignment like untold numbers of presentations I had done many times before. We put together our presentation and then divided it up, deciding who would lead the various parts of the presentation. When we completed our presentation, we were ready to do our "dry run."

There were three or four of us, and my section was close to the end, maybe even the closing section. When it came to my turn, my chest tightened up and I could hardly breathe. I felt an overwhelming sense of fear sweep over me. I could not speak. My friends asked me if I was ok, surely thinking I had suddenly become ill. I told them I had to go, but that I would be fine with my section of the presentation.

I had struggled with a fear of public speaking for all my life but had always been able to manage it. It had never been an issue with small groups and typically, my fear would subside within minutes, if not seconds. I had always been able to make effective presentations and public talks. Few people ever knew that I struggled with this, but this time was different.

I went to my hotel room, but the feeling of doom and fear did not subside. I was awake all night, overcome with fear and anxiety. I packed my suitcase to go home the next day just to be doing something. Morning came, I put on my suit and went to the lobby to meet my friend Mike for the ride to the training center.

Thank God, my group was scheduled to be one of the first groups to present. I met my teammates and, of course, they all wanted to know if I was feeling better. I told them I was fine, even though I wasn't.

We made our presentation, and we did well. We just wanted to pass and go home, especially me. I did fine and I do not believe anyone could sense the problem I was having.

I was so relieved. We had one more class session around noon and were then released to go home. I couldn't wait to see my family and for the nightmare of the past 18 hours to be over. But it would not be over.

This sense of fear and doom gripped my body and being for the next five years. Somehow, I was able to carry on. I think it was because I knew I had to. Except for Madelin, one close friend, God, and my doctors, no one ever knew about this. My kids didn't know about it until I told them after they were grown.

I covered it up well, but it was a tremendous burden and one that I felt I may never overcome. I could not have made it without Madelin's understanding and guidance and the grace of God.

My primary care doctor referred me to the Mayo Clinic to rule out any medical issues. They ran all sorts of tests and everything was normal. I was referred to a psychiatrist and went for counseling sessions for months. I enrolled in a Dale Carnegie course designed to increase confidence and public speaking skills. I completed the course, but like everything else, it was not much help.

Madelin and our family doctor were always encouraging me, assuring me that someday I would feel normal again. Still, I felt that I was in a tunnel of darkness that no one could understand, and I feared that light would never shine again.

After trying everything else, my doctor put me on Xanax. I believe it saved my career and possibly, my life. I was on Xanax for nearly five years – not just occasionally, but every day. I was able to carry on and feel somewhat normal. I was able to do my job, and more importantly, I was able to be a good dad and husband.

I was addicted, both physically and mentally, and I feared that my condition would return if I stopped taking it. My doctor gave me a taper schedule numerous times, but it was always too aggressive, and I could not do it. The withdrawal symptoms were similar to the symptoms of my original condition which made it even more difficult.

I was determined to get off the Xanax, but I had to do it my way. Finally, I began reducing my own dosage but just by a fraction of a tablet at a time. I kept the same schedule, but I took smaller and smaller dosages. I did that for a year. For the last few months, I was breaking the lowest dosage tablet into smaller and smaller crumbs, taking one small crumb four times a day.

Some five years after taking the first tablet, I took the last small broken crumb of Xanax. Except for rarely taking a Xanax tablet prior to a presentation or speech, I was off Xanax for good. Many of us have some cross to bear – this was mine.

Because of this condition, my first few years at DEC were difficult. Every phone call was a struggle, meetings were even worse, and presentations, both internal and client-facing, were my worst nightmares. But no one ever knew or sensed that anything was wrong with me.

I was determined to take advantage of working for this great company and determined to achieve success beyond anything I had done before. It was not good enough just to succeed, I was determined to be the best and to set the standard that my colleagues would want to follow.

I had always been that way to a degree, but for some reason, it was greatly exaggerated. I should add that this was an aspect of my personality that my psychiatrist wanted to talk about ad nauseam.

I worked harder during my DEC years than at any time in my career. I carried work home every night and rehearsed every meeting and every presentation in front of mirrors, and sometimes, in front of Madelin.

I subscribed to multiple newspapers and all sorts of trade publications, such as *Computerworld*, and read them every night. I read and studied every Gartner research paper that was written about DEC. I did my own "clip service" and kept every positive DEC article and every negative article about my competitors indexed in notebooks. I would not let Madelin throw away any of the *Computerworlds*. I kept them in boxes in our master closet, and she hated that.

My work obsession was obvious to everyone, but no one knew what was going on inside me. Looking back, I think it was because I was afraid that I could lose my career altogether.

At the end of my first year with DEC, I remember my boss, Bill Hughes, writing in my performance review that I was a "student of the business" – and I was. Our district manager sent me a handwritten note saying that he wished he could clone "Chuck Cliburns" across his entire district.

Phil Lowe, the president of Barnett Computing Company, the IT arm of Barnett Banks, once pulled my boss aside and told him that I was not just a salesman – I was a crusader.

That was intended as somewhat of a jab, but Bill Hughes thought it was the best compliment ever. He took the first opportunity to tell our team about it, letting everyone know that all our clients should view us as "crusaders." He always put me on somewhat of a pedestal, which I'm sure some of my colleagues grew tired of.

One of them once shared with me that Bill had told him in his performance review that he was no Chuck Cliburn. My colleague was good-spirited about it.

I say all this, not to boast, but to explain that the person everyone saw was not the person I was. I was not the confident and ambitious young man they saw every day. I was, instead, a person that feared that I may soon not be able to work at all, and maybe never again.

Chuck Cliburn

14

THE MISSION

I persevered and continued my mission of getting DEC into Barnett Bank, Blue Cross, and CSX. Blue Cross and CSX were as blue as blue could be, meaning deeply loyal to IBM.

While Barnett was very blue as well, there were parts of the organization that were not dominated by IBM – most notably, Barnett Operations Company (BOC). I soon found my way into Barnett through door number one and within a year or so virtually all my time was dedicated to Barnett. I still worked with CSX and Blue Cross but, except for some occasional departmental successes, my success was tied to Barnett.

Barnett had two independent technology companies, each headed by its own president. Barnett Computing Company (BCC), headed by Phil Lowe, was responsible for the transaction processing systems for the core banking functions. BCC was an IBM shop – as blue as they come. Everything, from the mainframes to the branch platforms and networks, was IBM.

Barnett Operations Company (BOC), headed by Reave Miles, was primarily a Unisys shop responsible for the check processing centers and other back-office functions. Phil Lowe and Reave Miles reported to Robert Mann, the Chief IT Executive at Bank Banks, Inc. – the holding company. BOC was my door number one.

I soon got to know Robert, Phil, and Reave quite well. We were not golfing or fishing buddies, but all three would take my meetings and listen to what I had to say. Eventually, Robert became somewhat of a mentor. I never knew why, but Robert was always particularly nice to me. I think maybe he just saw me as a young guy with potential and an uphill road at Barnett. He was old enough to be my dad.

He coached me on their internal company politics, especially on the relationship between BOC and BCC. There was competitive tension between these organizations. Robert once told me that the best way to get Reave to like something was to convince him that Phil did not like it – and vice versa. There seemed to be some truth in that.

Reave and his wife attended the same church that we did – Mandarin United Methodist Church. They were nice people.

Another BCC executive that I got to know well was Craig Chaires, the Executive Vice President of End-User Computing for Barnett Computing. In those days, there was a divide between end-user computing and core business systems. They just did not come together well – if at all. End-user-computing was generally PCs, LANS, word processing, email, and sometimes, small departmental systems. This was Craig's charter at BCC for all the Barnett subsidiary banks.

If there was a break from big blue in large IBM accounts, it was usually in end-user computing. It was not uncommon to find Wang Laboratories (Wang), Data General (DG), or DEC systems in these areas – mainly because IBM was not very good at it.

Not so much, however, with Barnett Computing Company. Except maybe for a rogue Wang word processor here or there or a few scattered Apple desktops for publishing or something similar, everything was IBM.

While IBM was the king of mainframes and large transaction processing systems, DEC was the king of end-user-computing – so this is where we attacked.

Robert had told me during one of our first meetings that Barnett, especially BCC, had a deep IBM culture that would be hard for us to break through. However, he added that it would be irresponsible to Barnett shareholders for them to not fully understand how DEC technology could help them be a better bank.

I know this message filtered down to Phil and Reave and their respective organizations, and it helped me immensely. I suspect that this was, at least in part, why Craig Chaires was always willing to meet with me and to support my sales efforts in BCC. Clear evidence, by the way, of the old sales axiom "sell high."

Craig was always cordial and willing to answer my questions. He explained the BCC organization to me, where all the IBM systems were, what they did, and even quite a bit about personalities, including who would be more receptive to listening to DEC.

He told me that I should get to know Chuck Bradley at BOC and Tim Matz at Barnett Banks, Inc. Chuck was the Senior Vice President of Operations for BOC and Tim was a top executive with the holding company – the head of retail banking. Chuck became a good friend, an occasional golfing buddy, and a solid DEC supporter.

I got to know Tim too, but not as well as Chuck and some of the other BOC executives. Tim became more involved a few years later when we made a major push to move their entire retail banking network to a DEC platform. He was not an IBMer and, like Robert, was open to new non-IBM options.

I met with Chuck just a few months after starting with DEC. He was an exceptionally nice guy and we got along well. He was primarily a "Unisys guy" in that he was responsible for item processing, which was all Unisys.

It was Chuck who told me Barnett Operations Company would soon be acquiring a company-wide integrated office system and that the project would be under the responsibility of Rod Black, their Chief Financial Officer. It could not have been better news as no one could touch DEC in office automation, especially Unisys and IBM.

BOC had its own IT organization that reported to Rod Black. Within a week or so, I was in Rod's office.

Rod was a young executive with an MBA and consulting background from his days at Arthur Andersen. He was self-confident with somewhat of a risk-taker spirit, but he fit the banker image to a tee. He always wore dark suits and starched white or light blue shirts with his collar always buttoned. But his sleeves were usually rolled up when his coat was off. He wore red ties and black banker shoes. Rod was all business but with a dry and quippy sense of humor. He loved to play golf and was very good.

We had a great first meeting. He confirmed that they would soon begin a major procurement for an enterprise office platform. While I didn't realize it at the time, Rod would become a true DEC champion and a good friend.

He introduced me to the IT Director and other members of the IT management team that would be listening to vendor presentations, putting together the RFP, and doing the evaluation. This would be a large contract.

For us to win, BOC would have to be willing to re-cable two separate office buildings into a single DECnet environment, implement a VAX/VMS computing platform that would be the first for Barnett, and implement DEC's ALL-IN-1 office automation application suite, also brand new to Barnett.

Unisys and IBM were well known to Barnett but that was about the only thing they had going for them. When it came to office automation, in my opinion, they sucked. In addition to IBM, Unisys, and DEC, the Barnett team also held initial discussions with a few other vendors but quickly eliminated them.

That was good news because almost anyone would have been a stronger competitor than either IBM or Unisys. In fact, if you wanted to eliminate two weak players in the integrated office automation space, IBM and Unisys would have been the first two to go in any objective evaluation.

I was at Barnett practically every day and spent a lot of time with the IT team continually taking them new material to read. I remember taking them a *Computerworld* magazine that was full of positive articles about the DEC solution we would be proposing. I highlighted the important sections of each article. I did the same for Unisys and IBM and put color-coded tabs on the pages so they could easily find them. DEC had about a 10-to-1 advantage over IBM just based on the number of articles. I am not sure that Unisys even had a single relevant article. If it did, it was probably bad.

The DEC articles were all positive and the IBM articles were mostly negative. I overwhelmed them with reviews from independent industry watchers like Gartner Group, Aberdeen, and others. It was easy to build our case. DEC was the king of midrange systems and office automation, while IBM and Unisys were terrible, and the entire industry knew it.

Before they created the RFP and started the formal procurement, they invited IBM, Unisys, and DEC to hold demos and presentations. I encouraged BOC to ask both IBM and Unisys if they used their proposed office systems internally knowing that neither would want to answer that question.

DEC used ALL-IN-1, DECnet, and VAX/VMS internally on our worldwide network – exactly what we would be proposing. I suggested that no vendor should propose a solution that they would not use themselves. They thought that was a valid point. I knew this could torpedo the day for both IBM and Unisys.

Our demo and presentation went great as they always did. We showed them everything they wanted to see on our internal production network. We were creating documents and printing them in remote locations, sending emails to DEC offices in other countries (not counties, countries), and showing them how we used VAXnotes internally – somewhat of an early version of Facebook, but for business. We showed them that we were running on a remote DEC VAXcluster and how the system would failover in the event of a system outage. In big-picture terms, we blew their socks off.

A couple of days later, they called me to ask that I stop by their office at the end of the day. They took me to a conference room, closed the door, and told tell me that the IBM and Unisys demos were both train wrecks. I knew then that we were going to win.

A few days later, they called again for another "closed-door" meeting. This time they asked me if I could provide some "supporting material" for the RFP. I took them hard copies of similar RFPs from other clients where we had won along with other similar stuff.

I met with them frequently as they were writing the RFP. By that point, issuing an RFP was not much more than a formality, but they needed to show that they had done a competitive procurement.

A month or two later, they announced that the award was going to us. I met with Reave Miles and Rod Black and both were excited about implementing the new DEC system. But they still advised me that Robert had to give the final approval for the contract. Robert soon approved the award and contract. We won!

While it was a large contract, the real significance was that we were breaking into a large IBM and Unisys account, especially a bank. It was a strategic win for DEC that was well-publicized throughout our company. It helped me establish a good reputation in the company and put the wind at my back for years to come.

The implementation went great with a few typical bumps along the way. I was soon part of the Barnett family and even had my own cubicle at Barnett. They loved the system and everything about DEC, including me.

Reave Miles, the rest of the executives, and most of the BOC employees now had DEC terminals on their desks and used the new system for just about everything. They became strong DEC advocates.

Reave agreed to do a video testimonial for us. DEC sent an A/V team to Jacksonville to do the video in Reave's office, and it was quite exciting. The video was used as part of DEC's corporate marketing material.

I became good friends with many of the BOC executives, especially Chuck Bradley, Rod Black, and the IT managers that reported to him. Chuck, Rod, and his team all loved to play golf, and so did I. We spent many days on the golf course together.

It was not unusual to get a morning phone call from one of the BOC guys that would go something like this:

"Cliburn, how does your afternoon look?" I always knew what that meant.

"Not bad," I would say.

"Good - Oakbridge at 1:30."

If you want to get to know someone, play golf with them a lot. I have always been perplexed as to why golf is not a mandatory course in business schools.

We continued to do well over the coming years at BOC. We won a significant deal in Barnet Trust Company too – somewhat of a blue bird sold by a company that specialized in bank trust systems that ran on DEC hardware. You have to learn to take the good with the bad.

15

THE KINGDOM

BOC was the logical entry point to the Barnett kingdom, but it only allowed us to go so far. To penetrate the IBM side of the kingdom would be a much more difficult task.

We were not a threat to their IBM mainframes – we could not compete there. The big prize for us would be to replace the IBM-based branch banking platforms with DEC technology and/or become the midrange standard for the Barnett enterprise.

It was accepted by many that our midrange technology was more advanced than anything IBM had to offer. IBM's midrange strategy and networking strategy were a hodgepodge of competing products and confused messaging.

Barnett's branch banking technology was old, and they were planning on replacing it. It would not be easy to play on IBM's turf, but we were determined to do so.

At the same time, we were working hard to expand our business in BOC and to nurture our relationships there. We were always selling at BOC, taking their executives to various DEC corporate events, playing golf with them, and making presentations on new DEC products. I even took Rod Black to play in the Digital Seniors Classic Pro-Am near Boston once.

In addition to being great customers, the BOC folks were great partners and always as helpful as they could be with our efforts in BCC.

Barnett also had numerous IBM midrange systems scattered throughout the organization, mostly in the affiliate banks and mostly IBM System 36s. That was a target for us too, along with the branch banking platform.

We were now into the late '80s, a transitional time for the computer industry and especially for IBM and DEC. Both companies were as cocky as ever at the field level, but none of us could see what was coming.

IBM still believed they should win everything just because they were IBM, and we still believed we knew the month and year that we would overtake IBM as the world's largest computer company.

Innovative companies like Intel, Microsoft, EMC, Oracle, Compaq, and Apple were circling like sharks and about to revolutionize the industry. The industry was changing, rejecting single-vendor integrated systems from companies like IBM and DEC. Clients were putting together multi-vendor systems, buying database software from Oracle, networks from Novell, desktops from Dell and Compaq, printers from HP, and so on.

IBM was still fending off mainframe clone manufacturers and falling victim to the monster they helped create – the corporate PC revolution. The BUNCH companies were just trying to survive and figure out which way was up. In fact,

Selling Information Technology

Burroughs and Univac, the BU in BUNCH, had already merged to form Unisys, in an arguably ill-conceived attempt to improve their collective fate. Their initial post-merger tagline was the *"power of 2."* Some of us modified it slightly to the *"square root of 2."*

Even though tough times were ahead, DEC was still riding high in the late '80s, and it seemed we could do no wrong, while IBM was becoming somewhat of the industry punching bag. Their local area networking (LAN) strategy was at odds with their traditional SNA networks, and their PC and midrange strategies were at odds with each other and with their mainframe cash cows. Many things in the IBM world did not work well together.

Between 1977 and 1988, IBM released five different midrange computer lines: the System/34, the System/36, the System/38, the IBM 9370, and the AS/400. None of these worked together.

IBM customers could not add one of the new systems to one of the older systems, like adding an additional locomotive to a train. Instead, they had to get rid of the entire train and replace it with a new train. Even worse, IBM customers could not move application software over to the new system and keep going. The software had to be converted, always a major task, or new software had to be acquired and installed.

It was my contention that IBM clients had to replace the entire train whether they needed to or not. And then, the new train would not run on the old tracks (the operating system and software applications).

DEC released the 32-bit VAX 11/780 in late 1977 and shipped the first VAX with DEC's VMS operating system soon afterward. VAX/VMS would be the flagship design of all DEC systems up until the early '90s. The VAX family of computer systems eventually ranged from single-user VAX workstations to VAX 9000 mainframe-class computer systems introduced in 1989. All these systems were compatible with each other and operated on the VMS operating system. Digital's VAXclustering technology allowed new VAX systems to operate alongside older Vax systems – like adding a new locomotive to the train.

That is how I positioned IBM's convoluted midrange strategy to Barnett and anyone else that would listen. Even the bluest of blue IBM zealots could not argue this point. Most of them had "been there and done that" and knew that I was right.

DEC customers of the '80s always knew what was coming next – a newer faster VAX with more VMS and DECnet features. IBM customers, to the contrary, never knew what was coming next or which way to go. Would the new 9370 "VAX Killer" be the way to go or maybe the new "Silverlake," i.e., the AS/400? They were, after all, different systems and completely different from whatever the IBM customer already had. It was a confusing story and one that DEC exploited to the fullest.

It was hard for BCC to ignore us, although, for the most part, they wanted to. But BOC loved DEC, Robert Mann was supportive of DEC, and DEC's midrange technology was clearly better than anything IBM could cobble together.

IBM was a master at mucking things up for its competitors by selling vaporware – things they did not actually have. They often leaked and/or announced a product long before they would actually "have it." Sometimes they would never "have it."

Selling Information Technology

It was like IBM throwing a lifeline to their customers that were about to drown in a sea of DEC's better stuff. BCC was always looking for the latest IBM lifeline and would grab at anything that IBM threw their way. This IBM phenomenon was not unique to Barnett Banks and is what inspired Digital's slogan and advertising tagline, "Digital Has It Now!"

One of the more classic examples of this was IBM's Systems Application Architecture (SAA). IBM announced SAA in 1987. It was a giant marketing splash. You would have thought IBM had discovered the cure for cancer.

DEC systems were application compatible from the smallest VAXstation to the largest VAXcluster, the industry loved that, and IBM was losing customers because of it. They were getting beat up constantly by industry watchers and the trade press because they did not have a similar capability. So, IBM announced SAA.

SAA was not a "thing." It was a vision of a "set of standards" that someday would supposedly allow applications to be written so that they could run across IBM PCs, midrange systems, and System/370 mainframes. But the "set of standards" did not exist.

Still, it created a huge buzz in the industry. Finally, IBM had countered the huge compatibility advantage offered by DEC. But not really. SAA was eventually quietly dropped by IBM before ever becoming of any practical use. Now, most IT people do not remember what it was.

But IBM was not run by dummies. The SAA announcement made a significant effect on the industry for a while.

Soon after its release, I was summoned to a meeting with BCC by one of the Barnett high-level and deep-blue managers. When I arrived for the meeting, I was escorted to a conference room to find several BCC managers waiting for me.

They were as excited as little kids on their way to Disney World, and they got right to the point. They wanted to know if DEC planned to make VMS "SAA compliant." I tried not to show my indignation, probably unsuccessfully. This had to be in the stupid question hall of fame. I would have bet my house that the question was planted by IBM.

I resisted the "surely you are not serious" response and instead told them that I would have to check and get back to them.

When I got back to my office, I called someone at corporate to at least go through the exercise to see how to best respond. After being told how stupid my customer was, I was instructed to tell them that DEC would be reviewing the SAA compliance standards once they were published and released and that IBM had not released a timeline for that. When that happened, DEC would take a position.

That is what I told them, and it never came up again. VMS never became SAA compliant nor did any of IBM's own systems so far as I know. This is what it was like trying to sell DEC stuff to the BCC side of Barnett.

16

ON THE HOOK FOR SADDLEBROOK

Once a year the Florida District held a several-day team building event that was attended by the entire Florida District. It was usually held at Dodgertown in Vero Beach or the Saddlebrook Resort near Tampa. It was a "mandatory fun" event.

Everyone was there, administrative people, field engineers, software support, and sales, probably 200 people or so. The idea, I suppose, was to build camaraderie and friendships. The different business units formed teams that competed in everything from volleyball to horseshoes. On the last night, there was a dinner with the different teams putting on skits – more competition. Even though I would have rather been home, it was fun, and it accomplished its objectives.

Still, DEC had to make sure we didn't have too much fun, so the first half of each day there were business meetings. This was typical corporate stuff, usually a corporate speaker, product presentations, and the like.

Around 1989 or so, my boss, Bill Hughes, called me a couple of months before the upcoming event. He told me that our District Manager, Jeff Hall, wanted me to make a 45-minute presentation on competing with IBM. While I did feel somewhat honored, I would have preferred to do a tightrope act for a circus.

Although I had been working for Bill for several years, he was not aware of my personal battle with public speaking, and I was not about to tell him now. I told Bill to thank Jeff for the opportunity and that I was on it.

My thoughts were immediately focused on making this a presentation that everyone would enjoy, remember, and find useful.

For days I thought about how to do this, mainly while I was driving around town. I seemed to do my best thinking that way. I came up with an idea, a theme, and a title, "The Market Perception of IBM and DEC." The presentation would be framed around this premise.

The idea was this. I believed that the best market perception an IBM product ever had in those days was on the day the product rumors began. Everyone wanted to believe that IBM was finally going to announce something that was actually great. This was particularly true in midrange systems.

Just like SAA, you could have made this argument with most any IBM midrange system. IBM's 9370 was touted as the "VAX killer," but mainly during the long period between the time it was only a rumor and the time it was announced by IBM. From reading industry trade publications, you would have thought that DEC would have had to close its doors on the day it was announced.

But when it was announced, the real reviews began with actual comparisons to DEC and other competitors. Was it a VAX killer? Not at all. The longer the 9370 went into its lifecycle, the worse its market perception became. I had press clippings and product reviews with timelines to prove my point. I presented this premise on graphs – a downward curve from rumor to product maturity – not just for the 9370 but for other IBM products as well.

The curve for DEC products, I asserted, was the exact opposite. The worst the market perception would ever be for a DEC product was on the day the product was announced. DEC under-promised and over-delivered. When DEC announced something, the industry would yawn.

This point held true with virtually any DEC product: the 32-bit VAX, VMS, DECnet, VAXclustering, ALL-IN-1, and so on. I had graphs for these too, with dates and documents to support my premise. These were graphs that went steadily up with time – the more mature the product became, the higher its perception became.

I put the presentation together and practiced over and over at home in front of a mirror. I even had Madelin be my practice audience from time to time. I knew I had a good and original presentation if I could just control my nerves.

Finally, the day came. We were at the Saddlebrook Resort in Tampa. I was quite nervous but comforted by the Xanax tablet I had in my shirt pocket which I strategically took an hour before I was on.

I was introduced and walked up to do my presentation. Within a minute or so, I was no longer nervous and was, instead, very confident in what I was saying. It may have been the best presentation I ever made. I knew that everyone was enjoying it and enthusiastically agreeing with everything I said.

Several of my colleagues came up to me afterward to tell me how much they enjoyed it, and I also received many complimentary emails in the following weeks. I ran into one of my old DEC friends in an airport years later and he mentioned that he still remembered my IBM presentation and told me how much he enjoyed it.

Successfully completing that presentation was one of my most treasured personal triumphs – right up there with conquering Spanish at Thunderbird.

17

A BRIDGE TOO FAR

Once that was over, I was again completely immersed in Barnett. The campaigns to become Barnett's "midrange standard" and to win the branch banking network went on for several years. Replacing IBM System 36s with DEC VAX systems as the midrange standard was one thing, but to replace the IBM branch banking network was a completely different thing.

Realistically, it was probably a bridge too far, but we persisted, dragging Barnett inch by inch. They objected to everything but never told us they were no longer interested, probably in large part due to the influence of Robert Mann.

For Barnett to change their branch banking system to DEC, they would have to change their wide-area network from IBM's SNA to DEC's DECnet, which would be a significant technical change and a monumental cultural change.

I came to know most of the BCC top managers quite well. I knew their personalities, their opinions and biases, and their circles of influence among their colleagues. A few of them became friends and were at least sympathetic to the DEC mission. Others were just plain obstinate. If I said it was a nice day, they would find a reason to argue that it was not a nice day.

The objector-in-chief was Charles Douter, a name that I always felt was fitting. He loved to hassle vendors, especially in formal settings. He was a constant pain in my side. But even though he would complain if you hanged him with the proverbial gold-plated rope, I always liked him.

Charles attended all the BCC vendor presentations, and some of the BOC presentations as well. I think BOC would sometimes invite him to be politically correct.

One such presentation/demo was at the Atlanta ACT. All the attendees, except Charles, were BOC executives. Charles was in full form, challenging just about everything we said and constantly interrupting. It was clear that he was annoying and embarrassing the BOC executives. A few days later, Reave Miles apologized for his behavior and told me that I should not take him too seriously.

BCC tried everything to dampen our enthusiasm and to quell our efforts to win their branch banking system, except to tell us to go away. They always left the door open just wide enough to encourage us to keep going.

They had plenty of real things to object to, but most of their objections bordered on the absurd. Finally, though, they thought they had us with a real one.

They told us that they were quite sure that they would be choosing Culverin as their new branch banking software platform and that Culverin only ran on IBM and NCR systems. Checkmate.

I was not a banking expert and had never heard of Culverin, so I immediately called Frank Giebutowski, one of our corporate banking industry experts with what I thought was bad news. He told me that Culverin and DEC had been working for quite a while on a partnering agreement and that Culverin would soon offer their software on the DEC VMS platform. He told me that the official announcement would be made very soon and that I could share this with Barnett. Uncheckmate.

Before calling the Barnett people that had told me this, I called Robert Mann. He told me that this was important news to them and that I should make sure that Phil Lowe and his team knew about this right away. He also suggested that I have the Culverin people meet with Phil and his team on this as soon as possible.

Within a week or two we had executives from Culverin and DEC's banking industry group in Jacksonville at Barnett's new Avenue's office complex to meet with the top BCC executives. This was now about convincing them that a DEC/Culverin solution would be the best of all worlds.

For hardware vendors, there was nothing better than having the preferred software vendor tell your prospect how great your hardware is, and this is exactly what Culverin did. We felt like we had pulled ourselves from the edge of the cliff yet again and that we had lived to fight another day.

While the branch banking effort was going on, we were also working to become Barnett's midrange standard. This project involved most of the same BCC executives and our champions from BOC. It was easier to build a compelling case for this because DEC was well-established as the technology leader in midrange systems, and we were already the midrange standard in BOC.

Barnett finally issued a loosely defined RFP of sorts for the new "midrange standard" after the IBM AS/400 had been released. The AS/400 had been one of the industry's worst-kept secrets of all time.

Silverlake, the pre-release code name for the AS/400, had been the buzz of the industry for a year or more before it was officially announced, probably because the IBM 9370, the VAX killer, had been such a dud. I remember Craig Chaires admitting that the 9370 was not going to work for them as the System 36 replacement, but that the Silverlake might be a good choice.

IBM was great at freezing the market by letting its customers fantasize about how great the next "IBM whatever" would be.

We had made numerous presentations and demos over a period of two years before Barnett finally asked for a midrange proposal. The RFP was vague with few specific questions, and it didn't have a formal scoring process. It would have been difficult to put together anything resembling a normal, objective, and quantitative RFP where the AS/400 would have outscored the DEC family of VAX computers, which is probably why they did it their way.

One of my DEC allies that was on the evaluation team told me that it was the consensus that DEC had better technology and a better solution but that some argued they should stay in the IBM family until IBM solidified their midrange strategy and product line.

They decided not to decide, agreeing instead to gather more information from DEC and IBM and to continue to kick the can down the road.

The IBM supporters on the committee would do anything to delay a decision until they had some sort of reasonable justification to select IBM. Still, we were positioned well for the midrange replacement project, making good progress on the branch banking project, and growing our presence in BOC.

Chuck Cliburn

18

WHEN YOUR CHAMPIONS LEAVE

Sometime in 1990, I received a phone call from one of the Barnett executives, probably Rod Black. Whoever it was, he told me that Robert Mann had just announced his retirement. It was a surprise to me and seemed to be to most everyone else as well. I saw Robert only once or twice after that. I really liked him and hoped that this was something he had planned for and that it was on his timetable.

From a selfish perspective, I was concerned about the impact this would have on my team's progress and the efforts we had put in Barnett over the previous few years. After all, Robert had been a good advocate for us, always doing what he could to make sure we were seriously considered by both BOC and BCC.

As I feared, things quickly changed for the worse. The midrange and branch banking evaluations were put on hold. All my meetings with Barnett now became speculation and rumor-mill meetings. Everyone had an opinion and a hope as to what was going to happen.

Reave Miles, the CEO of Barnett Operations, was younger than Phil Lowe, the CEO of Barnett Computing, and much friendlier to DEC. I had great hope that Reave would get Robert's job. There was quite the buzz that it might happen, but it didn't. My second hope was Tim Matz, with the holding company, but that didn't happen either.

Within months, Don Paigner, an outsider to Barnett, was named to take Robert's vacated position. Don had spent most of his career with a large northern bank. Phil Lowe retired as the president of BCC, and other BCC executives began to leave as well. I even received a phone call from Charles, my favorite Barnett nemesis, telling me he was leaving Barnett and looking for something new. It was a house cleaning at BCC for sure. Reave Miles eventually left BOC as well.

Don began to build his own team and brought in Rick Maples as his "go-to" person in BCC. He was, essentially, the new "Phil Lowe." I think Don and he had worked together previously.

It soon became clear that Rick was having nothing to do with DEC, and Don Paigner didn't want to talk about it. In one of my first meetings with Rick, he told me that he could not see any "value-add" in working with DEC. At that point, I had exceeded my Barnett patience threshold and my management chain felt the same way.

IBM had a legendary reputation of selling FUD (fear, uncertainty, and doubt) and of getting any detractors out of their way. Whether or not IBM had any influence on this massive shake-up at Barnett, I never knew and never will. Maybe IBM just got lucky. Or maybe not.

We all went through the motions for a while, but our long

quest to turn Barnett Computing Company into a large DEC account was over and we all knew it.

Although we had great success at BOC, BCC was the most devastating loss, or failure, of my career. I would soon take a new position with DEC.

Chuck Cliburn

19

THE UNTHINKABLE

Just as this was occurring at Barnett, the unthinkable was happening. IBM started losing money and lots of it. For 1992, IBM lost more than $1 billion, the largest corporate loss ever at that time. Between 1991 and 1993, IBM lost roughly $16 billion. And, even though DEC's profits and revenues had continued to grow through most of the 1980s, its fortunes had turned as well. In 1991, DEC had its first yearly loss. In 1992, Ken Olsen, DEC's visionary founder, and CEO retired. DEC was not doing much better than IBM

IBM would eventually right the ship, but it was the beginning of the end for DEC.

The reasons as to how IBM and DEC ended up this way are debatable. In my view, each took different paths to get to similar bad places.

IBM ceded too much technology to outside firms, like Microsoft, Intel, and others, and lost its technology edge. It also had too many overlapping, and arguably competing, divisions with incompatible visions and computer architectures.

DEC, on the other hand, had the opposite problem. It had

relied too much on the technical superiority of its own products including DECnet, VMS, and the VAX and Alpha chip technologies while ignoring the industry move to best-of-breed and "open" technologies. Mainly, DEC underestimated Intel and overestimated the market impact of its Alpha chip technology.

It seemed that overnight both IBM and DEC had gone from Wall Street darlings to corporate dinosaurs just trying to figure out how to change courses and stop the bleeding.

No-layoff policies came to screeching ends for both IBM and DEC. DEC reduced its workforce by 6,000 people in 1991 alone. It was a horrible environment. Everyone was always looking over their shoulder and it was relentless. We watched our colleagues get laid off – one by one.

DEC was acquired by Compaq in 1998, one year after Barnett was acquired by NationsBank. I guess the one thing you can depend on in the computer and banking industries is change.

20

SO, YOU WANT TO BE IN SALES?

During this period, I had the misfortune to serve as a coach and mentor for a well-intended but poorly executed career change workshop. DEC was laying off large numbers of field engineers, the people that installed, serviced, and repaired the computer systems. The idea was to offer some of them new careers in sales if they could pass the sales transition workshop.

This seemed to be a dumb idea out of the gate to most of us. DEC did not need new salespeople. We were laying off the ones we already had. Further, making a field engineer into a sales executive was somewhat akin to making an airplane mechanic into a test pilot. But no one wanted my opinion, they just wanted me to coach a team of six field engineers through the class.

There were about 30 field engineers in the class divided into five teams, each coached by a sales manager or sales executive. All of them would be laid off unless they transitioned to a sales job. To make this transition, they would have to pass the class. Most of them were terrified about being there and even more so about the likelihood of soon being without a job.

I arrived at the training center near Boston with the other coaches, the day before the students were to arrive, for a workshop on how this would work. The entire class was based on a simulated sales campaign to a large pharmaceutical company. The students would make sales calls on pretend executives of a pretend company. Each team would develop a sales strategy and a mini-proposal and then conclude with a formal presentation to the pretend evaluation team.

The pretend executives and evaluation team members were Digital instructors. The whole thing was over the top. The instructors wore costumes, including lab coats and fake glasses, and walked around carrying clipboards. They each were assigned specific personality traits: amiable person, analytical person, rude person, etc. They all overplayed their roles as if they were auditioning for a church play.

It was up to us, the coaches, to tell the students how this would work. For most of the class, the instructors remained "in character."

It was our job to help our teams with their assignments and to help prepare them each night for the tasks of the next day. It was horrible, none of these people had the slightest idea as to what they were doing. Unfortunately, they all knew that only a handful of them would pass and that everyone else would be terminated.

We also sat in as observers and note-takers for all the pretend sales calls. The training center had developed (or had bought) a structured sales process that the students were required to use. It had something to do with "hitting all the windows" – a process where the salesperson had to "hit all nine windows" on each sales call. The windows were things like making an introduction and thanking the client for the meeting, saying something personal and friendly like "Did you catch that fish?", telling the prospect why they were there, and so forth. The students got dinged for skipping a window and were graded for their overall performance. The coaches did not grade. We just took notes on the windows so we could defend our students and help them for the next day.

Finally, the last day came. That morning my team met, dressed in their suits and nervous as hell, to rehearse one last time for their final presentation. I was doing my best to keep them relaxed but with no success.

To my shock and dismay, as my team was still rehearsing, a few of their classmates stopped by to tell them goodbye. They were in tears after having completed their presentations. They had been told that they were not selected to move forward in sales and that they were being terminated. I was furious.

I immediately went to the head instructor to let him know what I thought. The idea of telling students that they had failed before the other students even had their chance to make their presentations was unprofessional, unfair, and just plain stupid.

I did my best to calm down my team, but there was no way to make chicken salad out of this chicken shit. I apologized to them for the unprofessional behavior of the training staff and told them that they all deserved better. I did my best to assure them all that they had bright futures, regardless of the outcome of this class.

I felt as if I had been a part of a mean-spirited corporate charade and felt badly about it for a long time. I still do. But I knew that the people I had coached appreciated my truthfulness, my interest in them over DEC, and the absence of corporate propaganda in everything I told them.

When I returned to my office, I sent a scathing email to the DEC training management team and copied my management team. I never heard much back other than the obligatory "thank you for your input – we'll look into it." I received thank you cards and emails from most of my team, most of whom had been laid off.

21

WELCOME TO TALLAHASSEE

It was not a good time to have your only account crash and burn. Fortunately, several of my good friends and colleagues from my Sperry days, including Darrell Wilson and Wayne Fountain, were now with DEC in Tallahassee. They were always telling me that I should come back to Tallahassee, probably so they could take my money playing golf.

The Branch Manager in Tallahassee was Steve Latimer whom they had apparently convinced that this would be a good idea. Steve was a likable guy and had done a great job in building Tallahassee into one of DEC's stronger Florida offices. When I would see Steve at various DEC meetings and functions, he would always tell me that he would love to have me come back to Tallahassee, even before the Barnett account crashed. He was a good manager with a particular talent for managing up the management chain.

Steve and I talked more seriously about the possibility. Although Madelin and I enjoyed living in Jacksonville, we had fond memories of Tallahassee. Brian would be changing schools to start his freshman year in high school and Jamie was still young enough that moving would not affect her all that much. We also had a fear of our kids having to learn to drive in Jacksonville.

Tallahassee had a strong DEC office and the chances of success there would be better than in Jacksonville. Additionally, Steve told me that he would get me approved for a corporate relocation package, a rarity for a lateral transfer. I reasoned that if the company paid over $100,000 to move us, they would not likely try to lay me off until everyone involved forgot about the $100,000 moving expense.

Steve got the transfer approved, including the relocation package, and we moved back to Tallahassee in August of 1992. I was 41. Twenty-nine years later, we are still here.

I joined the State Team as a Sales Executive and was assigned several state agencies that no one else wanted and where DEC had not been able to sell anything. But that was ok with me, I was glad to be in Tallahassee and to have a job.

Tallahassee was not exempt from the layoffs. Several of the salespeople were laid off within months of my arrival. I'm sure they felt that it was not fair for them to be laid off before me. But my calculation was right. They were not going to lay me off anytime soon.

It was nice to be back in Tallahassee. Even though it had been eight years, I had more friends in Tallahassee than I did in Jacksonville. Many of the state technology executives were still here and it was nice to reconnect with them.

During my first year back, the DEC sales rep in Montgomery, Alabama, left the company – probably having been laid off. DEC did not replace him and instead asked me to cover the State of Alabama while they figured out a better long-term plan. It was only a four-hour drive from Tallahassee, so it was doable.

Steve told me that there was a sizable deal in-flight there for a Legislative Management System that DEC should probably win. If there was any chance at all, it was better than anything I had going in Tallahassee, so I agreed to do it. I spent several months driving back and forth to Montgomery.

It was a real deal for sure and the State of Alabama guys made it clear to me that it was going to be awarded to DEC. Procurements were a little "less formal" in Alabama than in Florida in those days and, in this case, it worked to my benefit. It was a large deal, large enough for me to have a successful year and to make my quota.

The following year I managed to connect with Lorne Shackelford. Lorne was another ex-Unisys sales guy that had started his own PC sales firm, New Horizons, an authorized reseller of DEC PCs. New Horizons became my account.

Lorne was an excellent salesman, and he knew PCs. From my viewpoint, he knew more about selling PCs than DEC knew about manufacturing them.

DEC was not at the top of anyone's preferred PC list. Still, due to Lorne's sales skills and connections with the right people at the State, especially at the Department of Health and Rehabilitative Services, he sold tons of PCs and related equipment to my State accounts.

My routine was to go to Lorne's office once a month to make copies of all the purchase orders so that I could send them to the DEC order/entry police to get sales credit. While this was good enough to have a good year, my career had become stagnant and boring. This was not where I wanted to be 20 years into my career.

Further, DEC was declining fast, and everyone knew it. The enthusiasm for the new line of Alpha-based computers had faded before it even started. Optimism and pride had been replaced by dread and fear in the DEC workforce. The company was struggling for a cohesive message and a way to stop the bleeding. The once-mighty DEC would survive for only a few more years.

22

THE LIGHT RETURNS

Burroughs and Sperry had merged in 1986 to form Unisys. The new combination of companies had struggled mightily to become "one company" and to find the much-anticipated synergy of merging two large companies. Some, including me, felt that this synergy never came, but at least the turmoil of the merger was mostly over by 1994.

IBM had its problems as well and was trying to reinvent itself. Still, Unisys and IBM were the "big guns" in Tallahassee, and both seemed to have a renewed sense of direction and a good likelihood of long-term survival and success.

Since I had worked for almost 12 years with the two companies that formed Unisys, I knew them both well including their product lines and their very different cultures. Returning to Unisys was at least worth pursuing.

The most expeditious way to do that required a connection to Richard Gaddy. I knew who he was, but I did not now know him personally.

Richard was the top Unisys executive in Florida. He had been with the company for more than 20 years – all of it in Tallahassee. Everybody in the Tallahassee IT and state government worlds knew Richard Gaddy or at least knew who he was.

As a Vice President and District Manager, Richard was responsible for all public sector operations in Florida. There were four Unisys public sector branches in Florida that reported to him: State Government Health and Human Services, State Government General Government, Tampa Public Sector, and Miami Public Sector.

I felt that my best course of action would be to call Keith Stringfellow, my friend whom I had worked with at Burroughs in the mid-'70s. He had been a successful Account Executive in the Phoenix Named Accounts Branch and had moved steadily up the Burroughs/Unisys corporate ladder. He was now, some 20 years later, a senior Unisys executive. We had crossed paths a few times in Jacksonville since Barnett was a large Unisys account.

I called Keith at his Atlanta office to see if he could introduce me to Richard. Surprisingly, he picked up his phone, even though he had someone in his office. After a bit of catching up, I told him I was interested in possibly going back to Unisys and asked him if he would mind introducing me to Richard.

Some things are just meant to be. After putting me on the speakerphone he said, "Chuck, let me introduce you to Richard Gaddy." No kidding, Richard was in his office. What were the chances?

We were all laughing as Keith gave me a great introduction to Richard. Richard told me to call his administrative assistant,

Mary Lou Hunt, to set up a lunch meeting when he got back to Tallahassee. I called Mary Lou, a great lady, by the way, and scheduled a lunch meeting with Richard at Applebee's on the north side of Tallahassee.

We soon met for lunch and everything just clicked. Since we didn't know each other, we talked quite a bit about our backgrounds including our early days in the '70s with Burroughs. We talked a lot about the State of Florida, its IT direction, and some of the key state players. It was always fun to talk about them.

Richard told me about the Florida District and the Tallahassee office, how they were organized and who the key people were. He explained that Unisys was transitioning from a computer manufacturer to more of a services-led firm, adding that they had made significant progress, especially in networking and outsourcing.

We had similar sales philosophies and backgrounds and seemed to see eye to eye on just about everything.

I felt right away that we would work well together – and the feeling was mutual. At the end of our meeting, he told me that he was a big believer in chemistry in business relationships and that it was obvious that we had the right chemistry.

There was not an immediate opening, but he told me to be on standby and not to take anything else without calling him first. He said that there would be some changes soon that would open some things up. I left the meeting feeling good about the potential of returning to Unisys.

My good friend and colleague from Sperry and DEC, Darrell Wilson, had told me a few weeks before this that he would soon be leaving DEC, but that he couldn't tell me yet where he was going. Actually, he told me that he could tell me, but that then he would have to kill me. Within a few days after my interview with Richard, Darrell told me it was official – he was resigning and going to work for Unisys.

We enjoyed talking about this and the possibility of us working together again at Unisys. It seemed that we always followed each other in our careers.

Darrell had accepted a position in the Unisys consulting and services group and would be primarily responsible for vetting and developing large and complex services-led projects. My first response was "Congratulations!" and my second response was "Find me a job," and Darrell was all about it.

Within the next month or so, Richard called me twice regarding potential positions. This first one had no appeal at all, and the second one was not much better. The second one was for a new position they were creating for someone to lead PC sales in Florida Public Sector. I could not spell PC and it would have been a terrible fit.

Darrell still had his way of being "direct." He told me I would be an idiot, a specific type of idiot at that, if I did not at least talk to Unisys about this one.

I called Richard and told him that this position would probably not be where I could make my best contribution, but that I would at least be interested in learning more about it. He had the Unisys regional manager in charge of this program call me. Al Savageau was a seasoned Unisys executive and an extremely nice guy. Al called me and told me that he would be in Tallahassee soon and that we would get together for lunch.

My goal was to convince Al that I was a "must-have" but

not for this particular position. It was a tricky proposition, but that was the message I wanted Al to take back to Richard.

Al quickly figured out that I was not interested in this job, but he could see that I was very interested in Unisys, that I knew how to sell and manage, and that I knew Tallahassee. I think he would have hired me anyway if I had wanted the job. Sadly, Al died a few years ago – may God rest his soul.

It wasn't long until I got the third call from Richard. As they say, the third time is the charm. This time Richard told me the Branch Sales Manager of the Health and Human Services (HHS) Branch was leaving the company and that this would be the perfect position for me. I would report directly to Richard and manage the Florida Health and Human Services sales team. This team was responsible for all the HHS state agencies – the main one being the Department of Health and Rehabilitative Services (HRS). HRS was a significant Unisys mainframe client and bought all sorts of computer technology from Unisys. Now, we were talking.

I met with Richard again mainly to discuss the compensation plan, the sales team, and the business outlook (pipeline). It was all good. I then spent a day in Atlanta interviewing with Carl Bell, the Regional Vice President, Dave Williams another regional executive (and one of my future bosses), and Human Resources. Soon, I received the offer letter and accepted the position. It was the fall of 1994.

This marked the beginning of the most enjoyable 11 years of my career and a major career turning point as well. I was 43.

I was soon back at Unisys. My previous time with Burroughs and Sperry now counted as time with the company for my pension, vacation time, and consecutive years of quota attainment. All of these were beneficial "extras" that I was unaware of when I accepted the position.

Everyone was extremely gracious and welcoming. Still, I felt a sense of homesickness that lasted for a week or so. Fortunately, Darrell was there, as was another good friend from my Sperry days, George Zimmerman. I soon adapted to the new routine of getting up and going to the Unisys office instead of the DEC office and to the daily routine of my new job.

It had now been eight years since the Burroughs/Sperry merger. It was interesting to see how the companies had come together. The merger had been difficult, but everything was functioning again.

Many had been skeptical of a good outcome from the beginning. The two companies had been fierce competitors with aging and competing product lines, and the companies had fully duplicate organizations. It was not a friendly merger either, it was a hostile takeover with Burroughs being the aggressor.

The merger had temporarily made Unisys the world's second-largest computer company with over $10 billion in revenue, but that did not last long. By 1995, revenue was down to $6.3 billion with a continuing downward trend. Unisys revenue for 2019 was $3 billion, far less than either Sperry or Burroughs in 1986 prior to the merger. But Unisys still exists, which is more than one can say for DEC.

Unisys opted for the Sperry Univac headquarters facility in Blue Bell, Pennsylvania, to be the new Unisys world headquarters, but many, if not most, of the management team that was there in 1994 came from Burroughs. The Unisys salespeople in Florida were predominantly from the Burroughs side of the merger, as was Richard.

But there were some positive attributes from both firms. Unisys now had a strong customer support organization that, in my opinion, came culturally from Sperry and a stronger sales organization that came culturally from Burroughs. Gone was the Burroughs "Econo" approach to everything. Unisys now had modern offices with matching desks, cubicles, green plants, and great customer support.

When I started with Burroughs in 1974, it seemed that everyone was under 30. Now, it seemed that no one was under 40.

Unisys was a "big dog" among computer companies in Tallahassee in 1994 and was about to become an even "bigger dog."

Chuck Cliburn

23

BIG GAME HUNTING

Within a week or so after I started, Richard left for Africa for an elephant hunt. That was the first I knew of his love of hunting – especially big game hunting. He had some sort of permit for an elephant hunt and had been on a waiting list for an elephant that could be "hunted." When the call came, he had to go.

He returned in a month or so after a successful hunt and just in time for the start of the new business year. Richard asked me to help him plan and prepare for the annual district kick-off meeting. He told me the theme for the upcoming year would be "big game hunting." Imagine that.

I am not a hunter. I shot a bird with my dad's .22 rifle when I was about 10. I walked over to see the bird as it was lying on the ground, injured, and trembling, and looking up at me. Horrified and crying, I pointed the rifle at the bird and killed it. Other than an occasional poisonous snake, I cannot remember ever intentionally killing another animal. Madelin and I once trapped a rat in our house in a box. We took it out in the woods and let it go. I am not passing judgment on those who love to hunt, but it's just not for me.

Still, the "big game hunting" theme made sense. We were all hunters in a sense – always hunting for the next deal and the bigger the better. Richard wanted to carry the theme into our forecasts. The biggest deals would be elephants. The next would be water buffalos and so forth. We worked together to determine the size of the deals and the appropriate animal for each category.

Richard's idea was perfect timing for me. The company was all about finding "services-led" deals. These deals are characterized by higher risk, longer sales cycles, and long and complex implementations. I knew that I had several upcoming deals that would qualify for various big animals and one, for sure, would be an elephant. I was all in.

I liked the Unisys move to become more of a services company. Hardware had become a commodity and all the action was in systems integration and outsourcing deals. That was where I wanted to be.

Joining Unisys was one of the best things, probably "the" best thing, that ever happened to my career. I never sold another computer or any sort of computer hardware after joining Unisys. My team did, but it was largely without my assistance. My focus and energy for the rest of my career would be on large systems integration and outsourcing deals usually worth tens of millions of dollars and sometimes, hundreds of millions.

While I was still at DEC, Darrell had told me that he thought Unisys would probably go after the FLORIDA project at the Department of Health and Rehabilitative Services (HRS) adding that Unisys was not afraid to go after something big and "out of the box." He was right and the FLORIDA project was as big and "out of the box" as you could get.

FLORIDA was an acronym for *Florida Online Recipient, Information and Data Access*. It was Florida's Public Assistance and Child Support Enforcement system, and it was definitely "an elephant."

This project was essentially the mother of all staff augmentation projects. The FLORIDA system was a large and complex legacy COBOL/IMS system. HRS claimed that it was the largest IBM IMS system in the world based on transaction volume, and it probably was.

It was so large and so complex that HRS chose not to maintain this system on their own, opting, instead, to outsource it. At the time, it was outsourced to Deloitte, but the contract was near its end and it would soon be rebid.

HRS was also a mainframe account for Unisys, even though the FLORIDA system ran on IBM mainframes. As such, we had a team of salespeople, field engineers, and software support people that were, for the most part, dedicated to HRS. The sales team, led by Bob Burton, a great sales executive I should add, reported to me. The team was at HRS all the time and was part of the HRS family. We knew a lot about what was going on in this account.

One of the "things we knew" was that the C-level leadership of the agency, including the agency Secretary, had not been happy with Deloitte's performance on the FLORIDA project. However, the IT team that was responsible for the daily operations of the system and the management of the FLORIDA contract seemed happy enough with Deloitte.

We felt that most of the larger consulting firms that had public assistance qualifications (quals) would likely not bid due to the low rate structure that would be required to win. We also believed that traditional staff augmentation firms would not bid due to their lack of public assistance and project management quals and their inability to deliver such a large project. Even if they did bid, we didn't think they could win. We were right on both assumptions.

Prior to the RFP being released, our team spent months strategizing and convincing each other that we could win and deliver this huge project. We were all in. George Zimmerman was not as "enthusiastic" as the rest of us, but he was in. In fairness, George was a delivery guy, so it was his job to be skeptical about everything.

I put the deal in my forecast with a low probability and an amount much smaller than we all knew it would be. The further you got away from Tallahassee, the less enthusiasm there was, but even the most ardent order-preventers had a hard time saying no to the possibility of winning this project.

As to the size of the contract, Deloitte had more than 100 application programmers, business analysts, systems programmers, and trainers on the project. We knew it would easily be worth $50 million over a five-year term.

There were some good (ok, good enough) reasons for us to bid, but having the best solution or the best chance to win was not among them. The first "internal sale" we had to make was that, if we did win, we could deliver. We argued that we could and that it was not rocket science. Many of the personnel on the Deloitte contract were contractors and not Deloitte employees. Deloitte would likely have no use for them if we won, so our first step would be to transition as many of them as we could to Unisys.

So how could Unisys win?

Our main argument was that Deloitte may not win for any number of reasons. If they didn't win, we would be the "least worst" of all the other options. It was not the most compelling of win strategies, but you go with what you have. It was possible that Deloitte may bid but with much higher rates assuming they would have no serious competition. If that happened, our price advantage may overcome their technical advantage. Deloitte could be disqualified for not signing a form or for making some other fatal error, or Deloitte could simply decide not to bid.

I contended that there were only two things of certainty. One, the only sure way to lose was not to bid. And two, the executive leadership in the agency didn't like Deloitte.

The RFP was released in early 1995. It was the traditional RFP that the State used then but rarely uses now. It was rigid and would be scored by a team of evaluators using predetermined evaluation criteria and point weightings. The pricing would be scored by a point system as well. It was basically a proposal writing contest and there would be no negotiation.

The Unisys bid-review process was much less formal than it would become in the following years. Unisys was still a hardware company trying to become a consulting and professional services company. We would have never made it through the formal bid reviews of the coming years after all the consultant-types from Andersen, Deloitte, and Bearing Point people came on board.

Richard and George had done a good job in positioning this up the executive management chain. Still, it was a constant struggle with continuing objections, directives, and second-guessing up until the day we submitted the proposal.

This is where George's unique talent of disregarding everything the Blue Bell and Atlanta people said and Richard's ingenuity and Unisys experience came into play.

After we had all read the RFP, we knew the order-preventers would be emboldened and that we had to shore up our "we can do this" argument.

I remember sitting in our first internal meeting after we had read the RFP. One of the first questions someone asked was, "Who knows what Title IV is?" That pretty much summed up the challenge before us.

Most of the RFP questions were similar to this: "Describe your firm's approach to further automating and improving Florida's Title IV-D processes." None of us had a clue what the question even meant.

Someone quickly came up with the only hope. We would hire a firm that understood the subject matter to help us write the proposal and add them as a subcontractor so that we could claim that "Team Unisys" had experience with this type of work. I think it was Comsys (Cutler Williams) that we hired. I am pretty sure we hired Bill Barnett, a public assistance consultant, to help as well. If we didn't, we should have. Bill was a genius on this stuff.

We cobbled together a proposal. Our goals were to not get disqualified, to have at least an acceptable approach and solution, and to have a price lower than Deloitte.

Our proposal process was a low-budget and low-tech endeavor for sure. We were approved to develop the proposal locally, not using a Unisys proposal center. We borrowed a proposal manager from one of the proposal centers and used our local people and outside consultants to write and produce the proposal. Some of the best proposals I have ever seen came from Unisys, but this was not one of them.

After many late nights and an all-nighter on the last night for a few of us, we completed the proposal. Richard had been told by corporate that we could not deliver the proposal until someone called him with the "final approval" which we still did not have going into the last day.

On the day the proposal was due, Richard told Mary Lou not to answer any incoming calls until the proposal was in the hands of HRS. Unisys did not have a reliable email system at the time. Forgiveness was easier than permission.

Darrell and I delivered the proposals and went to the formal bid opening. When we walked into the HRS room, the Deloitte guys were already there with their mountain of boxes. They had already won the "number of boxes" and the "fancy box" categories. Deloitte and Unisys were the only bidders.

The waiting game was now on. Soon, all the corporate order-prevention calls changed from reasons not to bid to "When do you think we can close this?" From this point on, everyone from corporate was "here to help."

A month or two later I got a phone call around 5 p.m. from the IT Contracts and Procurement Director for HRS. In her mild and non-excitable voice, she said, "Chuck, are you sitting down?"

"Do I need to be?" I asked.

"Probably so," she said.

She told me that this was an unofficial heads-up that Unisys would be posted as the winner of the Florida PROJECT the next morning at 8 a.m. I'm sure I said, "You have got to be kidding me!" or something more colorful. She told me that we had lost on points but that we had the lower price by a small margin. The Secretary made the final decision to award to Unisys, overruling the recommendation of the evaluation team.

As soon as I hung up, I headed upstairs to Richard's office to give him and the other guys the unbelievably good news. When I walked in, there was a celebration in process with lots of high fives, "Hell, yeahs!", and perhaps an open bottle of some sort of beverage. HRS had called Richard too.

We celebrated before heading home and convened the next morning to start planning for the inevitable bid protest that we knew was headed our way.

For large State of Florida IT projects, you generally had to sell everything at least three times before it would stick. The first sale was getting the award. The second sale was withstanding the protest and keeping the agency from "spitting the hook." The third sale was getting the contract signed, which was never automatic. Sometimes, there were other sales steps to clear as well, as was the case with this one.

We immediately convened a conference call with our legal department and Rob Vezina, our outside legal counsel for procurement matters in Tallahassee. To put it mildly, it was unusual to have an agency go against the scoring for an award that was the result of an RFP. We lawyered up for the Deloitte legal onslaught.

Deloitte had beat us soundly on the technical merit section of the bid. Roughly speaking and going by memory, they outscored us by 100 points or more out of 700 points or so available. We gained a few points on the pricing section, but not nearly enough to overcome the points we lost on technical. Still, the agency Secretary made the award to Unisys "in the best interest of the State of Florida." The Secretary would have never done that without first clearing it the State of Florida legal people and the assurance that he had clear and sound legal justification.

Our attorneys reviewed the RFP, the relevant statutes, and case law. Everyone agreed that the agency was on solid legal ground. Rob was quite confident and believed it was somewhat of a no-brainer for the State to prevail. Deloitte, however, had quite a different view and filed their protest within the required period. The fight was on.

Protests are heard in Florida by an Administrative Law Judge (ALJ) within the Division of Administrative Hearings.

For large procurements, protests are quite common. Most of the major IT vendors that do business in Florida are all too accustomed to it. Rob Vezina had been there and done that many times.

We soon had our hearing date scheduled. The legal team worked closely with the HRS legal team as Richard and I and others worked with the agency Secretary and his team. We worked hard to assure them that they had made the right decision in hopes of keeping them from giving in to pressure and changing their minds.

There is always a threat that the agency would do that. Fortunately, the agency was dug in and sticking with their decision. Deloitte's decision to file and follow through with a protest only served to anger the top leadership of HRS, in my observation.

Soon, the hearing date came. The sales team was not at the hearing at the recommendation of our legal team. We all waited for the good news. But it was not good news. The administrative law judge ruled in favor of Deloitte.

Rob told us that the ALJ was simply wrong and that we should file an appeal with the District Court of Appeals. He was confident that we would win, probably on the first day. We were too far in to stop now. We filed the appeal and won on the first day just like Rob said. We won!

We soon signed the five-year elephant contract worth more than $50 million. This win changed my career and the direction of Unisys with the State of Florida. We were no longer viewed in Tallahassee as just a hardware company.

When the contract expired five years later, we bid again and won again for another five years for another $50 million contract. By then, the agency had changed its name to the Department of Children and Families (DCF). This time we were the incumbent with all the right answers and an exceptionally good Project Manager, Dan Bowman, who was loved by DCF. It was a much easier fight, but still a fight. Nothing comes easy in this business.

24

RICHARD GADDY

I loved everything about my job and all the people that I worked with, including the people in the Atlanta regional headquarters. They were actually "here to help" when you needed them but otherwise stayed out of the way. And Richard Gaddy was the best boss I ever had.

Richard and I worked well together and became close friends. He always sought my opinion and advice on business issues, even on things that were not directly related to my specific role. I have always appreciated the confidence that he had in me and everything that I learned from him. In many ways, it felt like my real job was Chief of Staff to Richard Gaddy, and that was fine with me.

I spent a lot of time over the years on "Richard runs." He loved to go to pawn shops, and he introduced me to this great adventure. It was not unusual to get a call from him in the middle of the afternoon. "You got time for a pawn shop run?" he would say. Other times it may be a "tractor run" to drive out in the country to look at something for his tractor – or a new tractor. We did a lot of "coast runs" too, sometimes to prepare his coast house for an incoming hurricane. And there was the ever-popular "Whataburger runs" that Richard, Bob Abernathy, George Zimmerman, and I still occasionally do.

On one of the "Richard runs" we started talking about our childhoods and there were similarities for sure. We both had humble beginnings growing up in the rural south in the '50s and '60s – Richard in North Carolina and I in Mississippi. We both married our high school girlfriends when we were 20. But much of Richard's story was like something from a Sunday night movie.

He told me that his dad had died when he was two years old. His mom, a beautician, would often tell him about his dad and how he was well known for his hunting skills, especially bird hunting. She told him that his dad was the best bird hunter in their county and was quite famous. Richard and his mom were poor, but she did her best to take care of him during his early childhood.

Eventually though, she had to tell Richard that she could no longer afford to give him the care that he needed and that she had arranged for him to live in a place with other children – a Methodist orphanage in Winston Salem. She told him that he would be able to come home for three weeks every summer and that they would be together on holidays like Christmas and Thanksgiving. Most importantly, she promised him that she would come to get him once she could afford to take care of him again.

When he was 6, his mom put him and his suitcase on a bus to go to the orphanage. He lived there until he was 11.

Richard told me that he didn't like it there and that some of the staff people were mean. He ran away three times but never got far. Despite all of that, he told me he had always been careful not to talk badly about the orphanage because they did their best to take care of him for five years.

He continued to tell me his story. Every Thanksgiving the orphanage would hand out slingshots for the boys to play with for a week. Somehow, Richard managed to get two and always turned in the "shitty one" as he described it. He hid the good one in the woods so that he could play with it all year. He taught himself how to hunt with his slingshot and could shoot a squirrel out of a tree and kill a rabbit on the run. He managed to teach himself how to clean the rabbits and squirrels and would sometimes cook them on an open fire in the woods.

Finally, some five years later, the day came that he had dreamed of. His mom came to get him. She told him that it would not be easy, but that he would never have to go back to the orphanage. She told him that he would have to work some to pay for his clothes, so he quickly got a paper route and later worked at a gas station and a cotton mill.

Richard said that he hadn't wanted to go to college but changed his mind when he was a high school senior working at the cotton mill. He explained that just about everybody in town worked at the mill and that it was backbreaking, nasty, and dangerous work.

One day he was on break with one of the mill workers that had worked there for all his adult life. The worker told Richard that he had been there for 20 years and that he was up to $1.15 per hour. Trying to be encouraging, he told Richard if he would stick with it, he could make a good living at the mill. Richard was making $1.00 per hour and still in high school.

Richard told me he immediately realized that, while he

didn't know what he wanted to do, he sure as hell knew what he didn't want to do.

His mom had saved $1,000 for him to go to college. That was a lot of money back then, especially for someone in her circumstance. That is exactly what he did.

I then understood Richard's love for hunting and his commitment to do his job, and everything he did, so well.

A few months later, he invited me to visit him on his new farm in southern Georgia about 45 minutes north of Tallahassee. It was just the two of us and it turned into something that we would do quite often over the coming years.

He needed a little help with a farm errand or two, but mainly he just wanted to show me his new farm and enjoy hanging out. And maybe talk some more.

The farm was beautiful with plenty of open pastures and wooded land as well. I'm pretty sure that the main functions of the farm were to provide Richard a place to hunt and fish, a place to do tractor things, and a place for a quiet retreat. It was perfect for all of those.

I drove up in my pickup on a Saturday morning and met him at his farm. He gave me the grand tour, showing me the various places where he was going to put deer stands and plant peas to attract deer. I kiddingly told him that was deer shooting, not deer hunting. He knew I was not a hunter and was always good-natured about me occasionally kidding him about hunting things.

I should expand on Richard's hunting. He was not your traditional southern hunter. He traveled the world on hunting expeditions from Alaska to South America to Africa and other places as well. He often spent weeks on foot with guides in some of the world's most remote places in pursuit of everything from bears and mountain sheep to lions and elephants. Like everything else Richard has done, he did it well and with deep passion.

So back to the farm. After the tour and errands, we settled in the rocking chairs on the front porch of the farmhouse. Richard offered me a "cold" beer. To Richard, it was always a "cold" beer even if it was not very cold. It's a southern thing.

Not long into our chat, he asked me if I knew about the loss of their daughter, Kelly. I told him that I did, that I was very sorry, and that I never felt right about bringing it up. He began to talk it about.

The first thing he said was that Barbara and he still cried a lot. I could see the pain on his face and hear the hesitancy in his unusually quiet voice.

He told me that she was beautiful, smart, kind, and athletic with a particular passion for running. Richard had a picture of her in his office, so I knew that she was beautiful, just like her mom.

He told me that Kelly had graduated from Auburn and had moved to Atlanta to begin her career with Arthur Andersen. She was happy, and they were all proud of her. As far as everyone knew, she was the picture of good health.

He told me that she was jogging on a spring day, as she loved to do, on a track in downtown Atlanta. She ran past two women and a little girl, then turned around to run backward so she could high-five the little girl, and then turned back and ran on ahead. When she reached the other side of the track, the women, and others, noticed that she had stopped and was kneeling. The women later told Richard and Barbara that they assumed she had lost a contact lens. But then Kelly collapsed.

Richard and Barbara received the call that every parent prays they will never get. Their beautiful and vibrant daughter was still alive but in extremely critical condition. Stunned, scared, and heartbroken, Richard and Barbara drove to the hospital in Atlanta as quickly as they could. After their agonizing four-hour drive, the doctor met them in the hallway before they reached Kelly's room and told them that Kelly's brain was non-responsive and that she would soon die.

On April 4, 1990, Richard and Barbara lost their beloved daughter, Kelly. May God rest Kelly Gaddy's soul and bless the many memories she left for her mom, dad, brother, and friends.

25

THE PATH OUT OF HARDWARE

In the mid-'90s, not long after I had returned, Unisys announced a major organizational change. Like most big tech firms, Unisys was abandoning its long-held geography-based management structure that was organized by branches, districts, regions, and so forth. Now the company would be organized by business lines. There would be three major Public Sector business units, and everyone would fall into one of these. These were: Global Industries (systems integration), Global Outsourcing and Infrastructure (mostly networking stuff), and Systems and Technology (mainframe computers and other hardware).

When the announcement was made, executive management did not tell us what our new jobs would be, and no one was happy about it. To all of us, it was a classic case of fixing something that was not broken.

The Branch Manager of the other Unisys branch in Tallahassee was Kevin Curry, my counterpart and colleague. Like me, Kevin was relatively new to his position but, unlike me, he and his wife, Karen, were new to Tallahassee. Kevin was ambitious, younger than I, and focused on moving up the corporate ladder. He was one of the most high-energy people I have ever known. Just being around him exhausted me.

I walked into his office shortly after the announcement, closed the door, and asked him what he thought of all of this. I knew we would commiserate in our displeasure.

He said something along the lines of, "This is bullshit!" and added that it was not what he had signed up for when he took the new job in Tallahassee. He had already put a call into Atlanta to see what was available and said that he was getting out of Tallahassee as soon as he could.

Kevin was good at everything, including managing his career. Soon, he was promoted to a VP-level management position in Atlanta and continued to move up the corporate ladder. He was always a great advocate for things that made sense.

In the coming years, Unisys would continue to transition to a services-led firm focusing primarily on systems integration and outsourcing. Larry Weinbach, the retired CEO of Andersen Worldwide became the Unisys CEO in 1997, further crystallizing the direction of the company. He was clearly a consulting guy and clearly not a hardware guy.

This was something that Unisys needed to do, and it was probably the only way out. Weinbach was there to lead Unisys in a new direction and find a path out of hardware. Even though many of us didn't see it at first, it was a step in the right direction.

The new organization was soon in place. The larger states, like Florida, were now headed by State Client Relationship Executives (CREs). These were VP-level positions that were responsible for managing the overall direction and strategy. In Unisys terms, the CREs were the "owners" of their respective states.

Richard Gaddy was named as the Florida Client Relationship Executive and I was named as the Florida Client Relation Manager (CRM) for Health and Human Services within Global Industries, responsible for systems integration and other services-led projects. It was essentially the same as my previous Branch Manager position but without responsibility for hardware. It was perfect for me.

Everyone in Tallahassee reported to someone that was not in Tallahassee. The organizational chart looked like a freeway map of Los Angeles. I reported to someone in Atlanta, but still had a dotted line to Richard. As far as most of us in Florida were concerned, Richard was our real boss, no matter what the organizational chart said.

Unisys continued to adjust to the transition to a services-led company, and Florida became an early shining star for Unisys Global Industries.

The FLORIDA project was a great success, reaching a total Unisys team of 185 people on the project. Unisys also won the contract to provide Florida's first digitized drivers' licenses, SACWIS/HomeSafeNet, and other services projects during this period. Our hardware and networking groups were doing well in Florida too. Unisys was the 800-pound gorilla in Tallahassee.

Chuck Cliburn

26

SOMETIMES YOU LOSE, SOMETIMES YOU WIN TWICE

Next in the line of elephants was Florida SACWIS. SACWIS is the acronym for Statewide Automated Child Welfare Information System. The Florida Department of Children and Families released an RFP for a new and modernized SACWIS in 1998.

It was released as a classic systems integration project where the vendor would take full responsibility for delivering the final project for a fixed price. This RFP marked the beginning of a project and spending debacle that became folklore in the Florida government and child welfare circles across the nation. It was a ten-year lesson in what not to do with a large complex State project.

Florida SACWIS was under my sales responsibility. We would win this $50 million-plus project not once, but twice.

Unlike the situation we had several years earlier with the FLORIDA project, Unisys was a national leader in the implementation of SACWIS projects. We had several recent state implementations that were all good references, and all were doing essentially what Florida was calling for in the RFP.

We convened a 20-person team to work on the proposal at our proposal center outside Washington, D.C. and worked night and day for more than a month. As usual with large proposal projects, the closer the deadline came, the longer the days became. For the last week or so, we were working 18-hour days. I worked on many excellent proposals during my career, but this proposal was probably the best of them all.

In our pursuit of proposal excellence, we decided to work one extra day knowing that we would have to transport the proposals to Tallahassee via a chartered plane on the last night before they were due. This was obviously bad practice for many reasons, but that is what we did.

We chartered a Lear jet, and my colleague, Mike Strange, and I returned to Tallahassee with the proposals the night before the proposals were due. We delivered the proposals to DCF the following morning. There were only two bidders – Unisys and Ciber (or one of Ciber's related companies, perhaps Metamor).

This time, we won the number of boxes contest and the fancy box contest. We were confident that we would outscore Ciber on technical merit but were quite sure that their price would be lower than our price of $59 million. It was possible, though not probable, that Ciber could make up enough points on price to overcome our point advantage on technical merit. Once again, the waiting game was on.

This time, there was no unofficial heads-up call from DCF the night before the posting. Today, all of this is done online but then it was still "old school." They posted the winner at the appointed time on a bulletin board in the DCF lobby.

Selling Information Technology

The posting was always an extremely intense event. It was like waiting for the jury foreman to read the verdict in a high-profile court case. There were probably 30 or so people in the lobby waiting to see who won – mostly people with Unisys and Ciber, subcontractors, and state people. At the exact appointed time, a DCF person came out, opened the glass door of the bulletin board, placed the posting with push pins, closed the door, and walked off.

I was there with several other Unisys colleagues in the back of the crowd. People moved up, took a look, and walked out. The Ciber folks saw it before we did. They walked past us on their way out without making eye contact. They looked as if they had lost their dog, which was a good sign for us. We won!

There is an exhilaration in winning large competitive deals that is hard to describe. It must be similar to winning the Super Bowl or World Series for professional athletes. Winning large deals advances careers, creates new jobs for some, and sometimes saves the jobs of others. Commissions and bonuses can be in the six figures for lead sales executives and managers and in the tens of thousands for others that worked on the deal.

Losing, on the other hand, is devastating. It can, and often does, cost people their jobs. There are no bonuses or commissions for second place. And unlike the Super Bowl, there is no paycheck or trophy for second place. Selling IT is not for the faint of heart.

We were soon back in the office celebrating and calling everyone to share the good news.

The next day I was at DCF for a meeting unrelated to the SACWIS win and ran into the DCF CIO in the hall. I stopped, said hello, and shook his hand thanking him for the confidence DCF had shown in Unisys by making the award to us. His reply was strange and not the typical "great job, you deserved it" type of response.

Instead, he told me that there were still some serious discussions that would have to take place and that we would be hearing from them to schedule a meeting with their executive team. Something was up, and it did not sound good.

We soon received a request from the Secretary's office to attend a meeting regarding the contract award. On the Unisys side, the meeting was attended by our legal counsel, a couple of people from corporate, Richard Gaddy, and me. On the State side, the meeting was attended by a half dozen or so top DCF officials. Irma Gruffy led the discussion for the State. I do not recall her exact position, but she held a high-level position at DCF.

This was one of the most bizarre meetings I was ever in with the State of Florida. I always found Florida officials to be polite and businesslike, even when they held what I viewed as unreasonable positions. This meeting, however, was an exception. Their position was absurdly unreasonable, and their main spokesperson was neither businesslike nor polite.

She seemed to find pleasure in telling us that our price was too high – roughly double Ciber's price. I think that maybe their strategy was to show no mercy, hoping that we would somehow quickly concede to their position. She demanded that we match Ciber's price while mistakenly casting blame on us for not knowing what their budget was.

We explained that we did, in fact, know what their budget

was, but that we were forced to choose between a solution that would meet their requirements or one that would meet their budget. We could not do both. We pointed out that Ciber had to make the same choice but had apparently taken the other option.

We told them that we would rework our solution with the goal of getting our price as close as we could to the Ciber price. This infuriated them even more, especially Irma. In an abrasive manner, she reminded us we could not change the solution, only the price. It seemed that she actually believed that we might do that. We remained polite, which was not easy, and told them we would get back to them soon.

Our debrief meeting back at the office was largely centered around the rudeness that we had just encountered and the absurdity of their position.

In the following days, we offered to lower our price from $59 million to $54 million, which did not fly. Soon, DCF withdrew the award and canceled the procurement. They had rejected the winning $54 million proposals for a state-of-the-art, end-to-end solution that met their own stringent and exhaustive requirements.

This decision proved to be a costly mistake for the State of Florida and a drama that would unfold over the coming ten years. It ended up costing several high-ranking Florida officials their jobs and became known, thanks to the press, as "Florida's $200 million debacle." Ironically, and somewhat poetically, it soon resulted in big wins for Unisys, Ciber-Metamor, IBM, and AMS-CGI. Just about everyone came out as a winner, except for the State of Florida.

Before DCF could reissue the bid, as they had indicated they were going to do, Jeb Bush was elected governor. This, of course, stopped everything. Governor Bush soon named a new DCF Secretary who, in turn, appointed a new CIO, Ronnie Houssel.

This transition took several months. Once Ronnie settled in, they decided not to re-release the original RFP after all, opting instead to be their own systems integrator.

In the opinion of many, this plan was not well-advised. In my opinion, it had about the same chance of succeeding as me building a functional spaceship and successfully landing on the moon. Realizing they did not have the requisite manpower or technical skills, they decided to release a new RFP for technical staffing that would work at the direction of the Department.

Essentially, this was a staff augmentation project. Under this approach, DCF selected two vendors to provide manpower for the project that would be billed on an hourly basis. There were no deliverables related to vendor payments. Once these technical people arrived at DCF, they did whatever DCF told them to do. The staffing vendors had no contractual responsibility for the overall success or failure of the project.

Ronnie was a true professional and a nice and bright man with deep IT and management credentials. But he was new to the State of Florida. I think he just overestimated DCF's ability to do something like this and the complexity of working with other Florida stakeholders, especially the Florida Legislature.

We did our best to talk DCF out of this idea. We even brought in one of our top SACWIS Project Managers from another State to meet with Ronnie one on one before the staffing RFP was released. We encouraged him to loosen their bid specifications and release another systems integration bid. Our advice was not well-received.

Soon, the staff augmentation RFP for HomeSafeNet, the new name for the SACWIS project, was released.

Even though we felt it was a bad idea for the State, we knew that we would have a good chance of winning the new staff augmentation contract. We had credibility with DCF from the FLORIDA project and we had significant SACWIS experience with other states. Knowing the magnitude of effort in front of DCF, we correctly believed this would become a very large contract over the years – something much larger than DCF envisioned.

We felt certain that the "cost savings" idea that seemed to fuel this concept was based on wishful thinking on DCF's part. On the positive side (for vendors), this approach put little responsibility on the vendors. The success or failure of HomeSafeNet would ultimately rest on the shoulders of DCF.

Just as before, Unisys and Ciber were the only bidders. This time, however, DCF awarded the contract to both firms. This kept DCF from being totally dependent on a single vendor for the staffing of the project.

DCF soon began issuing staffing requisitions to both firms for the required positions. We would submit qualified resumes to DCF. They would do the final interviews and select the candidates. Sometimes they would choose our candidates and other times they would choose Ciber's candidates. Over the years, I think there was essentially an even split between Unisys and Ciber.

This was an indefinite delivery/indefinite quantity (ID/IQ) contract which meant there was no predefined scope of service and no predefined contract value. We booked the sales and revenue amounts as the billing occurred.

This worked out great for me and others in the sales organization. I was quite good at convincing Unisys corporate management that the projected revenue stream for the coming year would be less than it always was. Without going into a lot of detail about how quotas and commission plans worked, this basically meant that we always exceeded our revenue and sales quotas. It was the gift that kept on giving. Year in and year out, the purchase orders kept coming and our revenue kept growing.

It seemed, for a while, that the project was going ok and better than we had expected. DCF held frequent public project status meetings, usually conducted by Ronnie. But as time went by, and as the state expenditures grew, scrutiny, especially from the Legislature, began to grow. It became increasingly difficult for DCF to defend their lack of progress and the huge amount of money that was being spent.

What began as concern by the Florida Legislature eventually became full-frontal panic and scrutiny. The press picked up on it too. DCF was in far too deep to stop the project, so the spending continued at the reluctance of the Legislature. At one point, and I suspect at the encouragement of the Legislature, DCF hired IBM to assume the project management function of the project. It was too late. While it may have helped some, it mainly increased the project expense.

Eventually, the Legislature was deeply involved in the oversight of the project, and DCF officials were testifying and explaining in front of legislative committees. There were a lot of numbers being thrown around as to the costs the State had incurred. By this time, the Legislature's number was $200 million. The project became fodder for Florida newspapers and government magazines with headlines like "The $230 Million Debacle" and "Computerized Tracking Project Years Behind and $198 Million Over Original Budget".

In late 2002, a new DCF Secretary was named, and Ronnie was forced out as CIO. He was replaced by a young executive from another state who was soon promoted to Deputy Secretary. Both new DCF leaders were under pressure to do something about HomeSafeNet.

Both new executives were skeptical about Unisys and our history with HomeSafeNet. We were guilty by association with the project.

We met with the DCF management team on multiple occasions to present our ideas on saving HomeSafeNet. Trying not to come across as "we told you so," we still made sure that they knew the real version of the project history, especially that Unisys had been selected in 1999 for an end-to-end SACWIS solution only to be rejected due to our $54 million price tag.

We made sure they knew that we had strongly discouraged DCF from being their own integrator. We argued that we were the only vendor that could save a substantial amount of the completed due to our knowledge of the work done so far and our SACWIS success in other States.

They were not having it. They loved to use the term "fresh start," which meant anybody but Unisys. I suspect that the term "fresh start" was stoked by another SACWIS provider and their lobbyists that wanted their shot at the project. But that's a guess.

DCF had also been told by a Florida legislator during a

committee meeting that he did not want to hear of any further SACWIS awards going to Unisys. Legislators usually do not know enough detail to make statements like this unless prompted by a lobbyist. Fair or not, we were cast with much of the blame for the failure of the project.

We did our best to overcome our guilt by association but to no avail. To no one's surprise, DCF eventually decided to give up on HomeSafeNet and awarded a new contract for a new system to AMS (CGI-AMS) in the mid- 2000s. I am not sure how much DCF finally spent with CGI-AMS, but according to a Florida Senate document published in 2007, DCF estimated the new contract with CGI-AMS to be $50 million. The HomeSafeNet name was dropped and replaced with Florida Safe Families Network (FSFN) for the new system.

Ironically and sadly, by 2005, DCF had spent tens of millions of dollars with Unisys on the HSN staffing contract, tens of millions with Ciber-Metamor, and millions with IBM. DCF had spent far more than the $54 million that DCF had turned down for a turnkey Unisys system and had nothing to show for it. Then, they spent tens of millions to start over. These numbers did not include the State's internal costs.

It was a classic case of what state governments should not do with a large complex project and a classic case of being penny-wise and pound-foolish.

27

MARRAKESH, MYSTERIOUS WAYS

During my career, Madelin and I were blessed to go on numerous company-sponsored trips to glamorous places around the world. Such reward trips are common in the IT industry and are intended to be rewards for the highest performers.

The most memorable of these was a Unisys trip to Marrakesh, Morocco, in April of 2000.

It was one of the most fun and interesting trips we ever made, and it was certainly to a part of the world that we would have never seen on our own. What made this so special was not the destination or the fun we had, but the amazing sequence of events that could only have been orchestrated by God.

There were at least four people from the Tallahassee office that qualified for the trip – a large number considering there were probably only 300 or so there from the entire company.

Dan Bowman, a close friend, and the Project Director for the FLORIDA Project was one of my colleagues that qualified for the trip. Dan, his wife Penny, Madelin, and I planned the trip so that we could travel together.

The company arranged for everyone to travel by commercial airlines to Paris and from there we took chartered jets to Morocco. The four of us decided to take advantage of the Paris connection and travel several days early so that we could spend some time there before heading on to Marrakesh.

Dan and I researched Paris hotels and randomly picked one that sounded good. We then chose our departure date and emailed all of this to the event people to have them book everything.

When we arrived in Paris, we took a cab to our hotel. It was all new to Madelin and me as neither one of us had ever been to Paris. The hotel was small and quaint and in the center of Paris.

When we got to the hotel, there was another couple in front of us checking in. They were speaking English and sure sounded liked Americans. One of us asked them where they were from. They replied that they were from Philadelphia on their way to a company trip and that they had come a few days early to spend some time in Paris. Yes, they were with Unisys and headed to Marrakesh. We all made introductions and went to our rooms.

For the next few days, Dan, Penny, Madelin, and I did the usual Paris things – the Eiffel Tower, the Louvre, the Notre-Dame Cathedral, various outside cafes, and so forth. One of the days we took the "Chunnel" to London. For me, that was the most fun day. I sometimes say that my favorite day in Paris was the day I went to London. But it was all fun, especially being with good friends.

A few days later we were on our charter jet with other Unisys people from all over the world headed to Marrakesh. As usual for such events, we were met by charter buses and taken to an upscale resort. We quickly observed that people who live in Morocco are either Bentley-wealthy or donkey-poor. There was little in between.

Dan and I had both signed up to play golf the next day. We belonged to the same club in Tallahassee and played golf together all the time. Naturally, we were assuming, and hoping, that we would be on the same golf cart or at least in the same foursome. However, the event planners had already assigned the carts and the foursomes.

I found my cart and I was not with Dan. Instead, I was riding with someone named David Penn, which was fine as I have always been fine playing with strangers. Soon, David walked up with clubs. I had forgotten his name, but he was the guy we met while checking in the hotel in Paris. We had quite the chuckle about ending up playing golf on the same day, in the same foursome, and riding in the same cart. He was nice, and we enjoyed our round of golf. He told me that it was his wife that worked for Unisys.

For the next few days, we did all the "mandatory fun" things including taking buses one night to Casa Blanca for dinner. It was an extraordinary trip for sure. Unisys even hired America, one of my favorite bands, for a private concert and it was incredible.

The main business meeting was held on the third day. They told us not to miss it and to bring our spouses as we would have a keynote speaker that we would remember forever. I was skeptical of that.

Usually, the keynote speaker would be an analyst with an

industry-watcher like Gartner or something similarly boring. Madelin had heard a few of those speeches and had no intention of hearing another one. Somehow, I convinced her to go with me on the outside chance it may be someone interesting. She warned me that if it wasn't, she was going to get up and go back to the room.

As soon as our CEO began introducing the keynote speaker, it was obvious that he was talking about Mikhail Gorbachev. We were whispering to each other, "This is a Saturday Night Live spoof."

Finally, he said, "Please welcome Mikhail Gorbachev." From the back of the room, came Mikhail Gorbachev, flanked by six or so big muscular men in suits – bodyguards we guessed. We were near the front, not more than 20 or 30 feet away.

He spoke (through an interpreter) about the cold war and various world events, but mostly, he spoke about Ronald Reagan and their relationship and times together. He talked about Reagan's "Mr. Gorbachev, tear down this wall" speech. It was one of the most interesting speeches I had ever heard. Madelin and I were both happy that she went with me to the meeting.

Despite Gorbachev, America, and having dinner in Casa Blanca, there was still something coming that would define this trip for me and be an important life event for someone else.

The customary formal banquet was held on the last night. I never cared much for these – they were all the same. It generally took forever, the food was bad, and the speakers were boring. Ditto for this one. As always, the banquet hall was full of large round tables with assigned seating, so the first order of the evening was finding our table.

We found our table, sat down and who do you suppose we were sitting with? David and Mary Penn, the couple we met while checking into to the Paris hotel, and the man who was my golf cart partner. There were two other couples seated at the table as well.

This became the first topic of conversation for our table. We were laughing about it and making jokes about this being some sort of weird karma. We told the other people at the table about this strange sequence of events and that Madelin and I were from Tallahassee and the Penns were from Philadelphia.

We explained how we had met in Paris, that we all had randomly selected the same hotel, traveled on the same day and arrived at the check-in desk at the same exact time. Then, David Penn and I had chosen to play golf on the same day and were randomly assigned to ride on the same golf cart. And now, we were sitting together in assigned seats for the banquet. Not only were we at the same table, Madelin and I were sitting next to David and Mary. I was sitting between Madelin and Mary.

The dinner dragged on forever. You know, 45 minutes for someone to come ask you what salad dressing you wanted and if you preferred fish or rubber chicken, another 45 minutes to get the salads, and so on. There was plenty of time for everyone to talk and get to know each other.

With 300 people or so in the room and with the tables being so big, it was hard to hear anyone except the people you were sitting next to, so most of my talking was with Mary and Madelin.

At some point during dinner, I asked Mary if she came from the Burroughs side or the Sperry side as most everyone was there before the merger. She told me she came from Burroughs and that she hated Sperry Univac with every fiber of her being. Then she began to tell me her story.

She told me that her dad had worked as an Account Executive for Sperry Univac years ago. He had been very successful, and everyone had loved him. She talked about him in glowing and loving terms. It was clear that she had a special bond with her dad.

She told me that some 20 years earlier, while she was still in high school, she stopped by her dad's office late one afternoon to say hello and to talk about their family vacation that was coming up the next week. She said she hardly ever went to her dad's office, but on this day she did.

She said she had been worried about her dad as he had been under a lot of pressure at work. She had hoped the vacation would be good for him.

She went on but with sadness on her face. While she was in her dad's office his boss came in, whom she mentioned by name. In trying to be polite and to engage in a conversation with him, she told her dad's boss that they were talking about their vacation coming up the next week. She paused as tears welled up in her eyes. I could sense the rage inside her.

With a tremble in her voice, she said her dad's boss told her that he was not going anywhere until he was back on his numbers. He then turned and walked out. She said her dad was humiliated and heartbroken that she had to hear this. Like any good dad, he tried to assure her that everything would be ok.

With tears coming down her face, she said that her dad had a heart attack and died in his office the following day.

While she was still in high school, it became her goal to

someday work for Sperry's most hated competitor even though she did not know who that was at the time. She wanted to spend her career making life miserable for Sperry. "That's why I went to work for Burroughs," she said.

By now everyone at the table was riveted to what she was saying.

I knew her dad. "Your dad was Fran Ashburn, and he was a great man," I said. She was crying and I was choked up too. I told her how I had worked with him in the Philadelphia Sperry office in 1979 and early 1980, and how he was admired and loved by all of us. I told her that we all viewed him as somewhat of the elder statesman of the office and that when we needed advice or just a friendly chat, we would talk to Fran.

Crying even more, she said, "Thank you" and left the table with her husband, David, following her.

Fran had died a few months after we had moved from Philadelphia to Tallahassee. Someone had called me to tell me about it shortly after he died. It had been 20 years since this happened.

I knew him well, and I knew his boss. He was my boss too.

David came back to our table after about 30 minutes and told us that she had gone to their room. He thanked me for listening to her story and told me that she rarely discussed her dad's death, especially with strangers.

He said that she had been waiting for a long time to hear some of the things that I said. I believe it was final confirmation, and perhaps a missing one, that her dad was loved and respected by his co-workers.

All of us at the table knew that this unlikely sequence of events was far more than a coincidence. May God rest the soul of Fran Ashburn and bless the memories he left for his family and friends.

28

THE STATE TECHNOLOGY OFFICE (STO)

During my years working with the State of Florida, the IT governance structure has, in my view, changed far too many times and not by just moving around the boxes in the organization chart, mind you, but by legislatively terminating (and/or defunding) one agency in favor of creating a new one. For those like me who have been in Tallahassee for decades, it has been like watching the same movie over and over. Four such do-overs have occurred starting with the STO in 2000.

This exercise in history-repeating never seemed to be of particular concern to most of our elected officials or the appointed heads of state agencies. This is probably due, at least in part, to term limits for legislators and short tenures for agency heads. Most of them, understandably, do not know the history of Florida's IT governance and could not begin to name the previous IT agencies or offer an opinion as to what has gone wrong.

In my opinion, except for the first attempt under Governor

Jeb Bush, Florida has never given the IT agency *du jour* the authority to do what needs to be done. In fact, until Bush's State Technology Office (STO), Florida never attempted to have a state CIO.

Before the STO, most agencies had their own IT division, their own data center, and their own CIO. Each operated independently with little knowledge or consideration of what the other agencies were doing. This governance model had many problems and had clearly outlived its usefulness.

Since the STO, each subsequent iteration of the statewide IT agency has been heralded as some sort of bold new direction, but timid new directions would be a better description. Only the STO qualified as bold. Governor Bush had the right idea, but poor (ok, horrible), execution. The STO did not just fail. It failed big. This colossal failure encouraged the agencies to continue with the status quo and crippled the Legislature's will to be bold again.

While Florida does have a consolidated data center now, most of the real control of Florida's IT systems remains with the agencies as they have full responsibility for their own applications. This makes it unnecessarily difficult for Florida to implement systems at an enterprise (cross-agency) level.

Since the demise of the STO, the Legislature has remained reluctant to give any significant power and authority to any version of a statewide IT agency. Instead, each iteration has been given varying degrees of limited authority. Further, these various iterations have operated without having their own budget, operating instead under somewhat of a "chargeback" system with the agencies. This has hampered real success as well, in my opinion.

The latest, and current, iteration, the Florida Digital

Service, went into effect on July 1, 2020, so the jury is still out on this one. While the new leadership is determined and qualified to provide a new vision and leadership, it still lacks the legislative authority to be truly effective in my view. Maybe more authority will come with time.

So, what happened with the State Technology Office?

Established under the Bush administration in 2000, the STO was the State's first centralized organization for statewide IT policy and enterprise operations. It was, indeed, a bold change in direction for the State of Florida.

Governor Bush named a previous associate of his, Rory Coles, to be Florida's first CIO. Soon after his appointment, Rory held a town hall meeting at the Leon County Civic Center and did a good job of introducing himself and the new agency and outlining his vision for the STO. Still, most of the agencies and vendors did not quite know what to make of this.

None of us knew Rory, but after a while, Richard and I met with him for dinner at the Tree Steak House in Tallahassee. Our main objectives were to get to know him and to learn as much as we could about his plans, especially those that may relate to our existing contracts and future business. We learned that he was big in stature, amiable, and technically astute. Other than that, we left the meeting not knowing much more than we did to start with.

We did our best to position Unisys as a partner and to deliver the message that we wanted to help where we could. We did not have much success, but in fairness, I am sure he was hearing the same pitch from everybody.

For a while, the State continued to operate on a status quo basis, and nothing seemed to change much. Less than two years later, Rory resigned and Joy Banes, one of the STO managers, took over as acting CIO. She was subsequently appointed CIO effective, July 1, 2002.

Under Joy, the STO developed somewhat of a bunker mentality and a "we are the STO and you are not" attitude. It became increasingly difficult for us to even get meetings.

We did our best to develop a relationship with Joy and her staff, but we had little success. We even brought in one of our woman executives to meet with Joy one on one, hoping that would help. Joy took the meeting, but it didn't help much.

There were rumors of big changes coming from the STO but no real signs of anything definitive. But the rumors were correct – big things were coming.

Just a few days before Christmas of 2002, I stopped by the Department of Children and Families to say hello and to wish everyone a Merry Christmas. Several DCF folks were there. One of them asked me if we were going to bid on the "STO ITN." It was an "oh, shit" moment for me, for sure. I had no knowledge of a "STO ITN."

The ITN in question was ITN 006. It had been released on December 20 and was one of the most significant IT-related bids ever released by the State of Florida.

The implications of this bid were gargantuan to both the vendor community and the state agencies. The State would be outsourcing a wide range of functions from security to data center consolidation, all on an enterprise basis and probably all to a single vendor.

From a vendor perspective, if there is anything worse than losing a large deal, it is not knowing about an important bid that is on the street. This entire STO fiasco was not a good thing for Unisys and was not helpful to my career. But for the winning vendors and the State of Florida, it was worse.

Selling Information Technology

This was one of those rare instances where not bidding, or losing, turned out to be better than winning. In the not-so-distant future, we would look back and say, "Thank God we didn't bid."

There were all sorts of things wrong with this procurement that would later be well-documented in the press and in a State of Florida Auditor General's report *(Report No. 2005-008, July 2004)*.

I was always skeptical of major bids that were released just before Christmas with short turnaround times. Such bids can be intended to be "under the radar" and difficult for vendors to respond to. For whatever reason, this is what the STO did with one of the most important IT bids ever released by the State of Florida.

Projects of this size (this one was projected to be worth more than $300 million) were typically preceded by detailed feasibility studies, cost estimates, savings projections, and market analyses – but this one was not.

The lack of customary pre-bid work could have occurred for any number of reasons. Regardless, this bid fell short in this area and the STO was called out for it by Florida's Auditor General.

Tallahassee is a small town with a tight-knit vendor community – we all knew each other. It seemed to most of us that BearingPoint probably had the best chance of winning and that Accenture had the second-best chance. We were quite close on that call.

The proposals were due in early February. With the expansive scope of the ITN, our lack of previous knowledge of the deal, the poorly defined requirements, and the short turnaround time, we did not have the time or the desire to submit a proposal. Our only option was to subcontract with BearingPoint or Accenture and hope that the whole thing would blow up. And it did with a seismic boom.

Richard and I, and others, met with the local leadership of both Accenture and BearingPoint, hoping to establish a subcontracting relationship. The Accenture meeting went nowhere. They did not see a need to have us on their team and they were probably right.

The BearingPoint meeting went better, but not much better. At least they were willing to have a couple of meetings with us to discuss the possibility. There were at least two formal meetings where we presented our not-so-compelling case and listened to what BearingPoint had to say.

From our meetings, it was obvious that they had a solid knowledge of the bid and a clear vision of their solution. This was particularly "impressive" given that the ITN had only been out for a week or two.

BearingPoint seemed to be on a first-name basis with just about everybody at the STO. There were a lot of "Joy said this," and "Joy said that" comments by their team. One of the BearingPoint executives received a phone call or two from Joy during these meetings. We found that to be quite "impressive" as well.

According to the AG report, there were 39 proposals received with 19 being deemed as "lacking sufficient information" to be evaluated. BearingPoint and Accenture both submitted qualifying proposals.

The ITN process in Florida gives the agencies significant latitude in the procurement process. The agencies can, to an extent, do whatever they want. They can choose one vendor with which to negotiate, or they can negotiate with multiple vendors. They can propose a post-proposal "shotgun marriage" approach where one vendor does one thing, and another vendor does something else, assuming the vendors agree. In the end, they award to whomever they determine to have the "best value" for the State. Every ITN process is somewhat different, but this one was off the "different scale."

There was a general feeling in the vendor community, and even among some of the people with the State, that this procurement had not been particularly, let's say, "well-managed". There was suspicion among some, including us, that BearingPoint may have somehow influenced the ITN, and perhaps, had prior knowledge as to when the ITN would be released. It was quite the rumble in the vendor community.

On February 28, 2003, the STO issued its intent to award to BearingPoint and everyone did not live happily ever after.

Even the issuance of the intent to award was "unusual." With an ITN, after the proposals are evaluated and scored, the State would generally post the vendor, or vendors, with whom they would negotiate. Then, after negotiating with at least one, the agency would post the intended award. In this case, the STO skipped the negotiation step, or so they tried and went straight to posting the award for BearingPoint.

Accenture and others immediately filed intents to protest. Then, the STO filed an "after the fact" Intent to Negotiate which nullified the Notice of Intent to Award to BearingPoint. This triggered an intent to protest by BearingPoint for the STO reneging on the original intent to award to them.

It was a procurement goat rodeo that had to be pure hell for the STO and the entwined vendors. But for the rest of us, it at least had some redeeming entertainment value.

There could be no good outcome with any sort of protest

that went to the end. It would only expose the faults of the procurement, delay (at best) any award, and most likely result in the entire procurement being thrown out.

Then there was the MyFlorida Alliance. According to the AG report, *"(The) Alliance was born out of a subsequent undocumented mediation process between the STO and protesting vendors."* As a result of this "undocumented mediation," contracts were signed with BearingPoint and Accenture on August 13, 2003. That way, the STO, Accenture, and BearingPoint all won. The STO saved face (for a short while) and Accenture and BearingPoint divided up the pie.

The whole thing soon started to unravel. There was too much visibility and too many lingering questions about the entire ITN 006 process that would not go away.

Joy Banes resigned in February of 2004 only months after the contracts were signed and took a position with BearingPoint. A highly regarded Deputy Chief of Staff for Governor Bush took over as CIO a couple of months later.

In July, the aforementioned blistering report regarding the procurement was released by the Florida Auditor General. Among other things, the report found that the STO did not adequately document the decision to outsource, nor did they adequately and properly evaluate the proposals. This was the beginning of the end of the STO's short life.

In October, the new CIO announced that all contracts related to the MyFlorida Alliance would be terminated effective December 30. And just like that, as a famous person once said, the contracts were gone and the STO was not far behind.

In what would come to be a recurring process for the State of Florida over the next two decades, the Legislature sought to terminate the STO via legislation, but the Governor vetoed the bill. However, the Legislature defunded most of the key positions and others were transferred to the Department of Management Services. The STO was later terminated by legislation.

Chuck Cliburn

29

SPACE INVADERS

It was now 2004. Unisys remained committed to the idea of transforming into a consulting and services-led company. We wanted to be Accenture or Deloitte Consulting when we grew up, and I loved it. I loved the large and complex deals, the strategy, the politics, and my Unisys colleagues. But there was a growing culture in Unisys that did not bode well for many of us that were Unisys veterans.

The CEO of Unisys, Larry Weinbach, had come from Andersen (Accenture) and the head of Unisys Global Industries (GI), Greg Baroni, came later from KPMG. The "consulting culture" was taking over.

Unisys was committed to the idea of transforming Global Industries into a consulting and services organization in the mold of the "Accentures of the world" – even changing the titles of top consulting positions to "Partner." It made us sound more like Accenture.

Everyone in GI that was not a recent hire from a consulting firm became unofficially referred to as part of the "legacy organization" and that was not a good thing. This term was used regularly and openly in the company. People would often use the term in describing someone. "Joe Smith is the legacy guy in Memphis" or "The entire Tallahassee operation is part of the legacy organization."

Then came the seller-doer model. In the seller-doer model, the same people that sold projects delivered the projects. The legacy organization did not have seller-doers. We had sellers, and we had doers, but not seller-doers. The seller-doer model was commonly used in IT consulting firms.

The GI organization was hiring every seller-doer they could find. The new seller-doers were everywhere coming from KMPG, Ernst & Young, Accenture, Deloitte, and other consulting firms. Many of them were hired as "Partners."

The seller-doers started showing up in Tallahassee for no apparent reason other than "being here to help." It was like getting calls from telemarketers. "Hi, Chuck, I'm the new partner responsible for blah, blah, blah." They all wanted to come to Tallahassee to meet with our team to see how they could "help." We heard it constantly.

The seller-doers, at least those that came to Tallahassee, did not impress me as particularly good sellers or particularly good doers. At least, legacy guys were good at one or the other. But that was my opinion. Maybe they were great at selling and doing.

These folks were always "here to help" but never demonstrated any specific helpfulness that I could see. As they say, I might have been born at night, but it wasn't last night. It seemed that maybe they were here to observe and to report back to the seller-doer mothership.

I was never on board with the new model – the same for many of us in the "legacy" organization. The reverse seemed even more true. The seller-doer advocates were not on board with us. I think that all the success Florida had had with separate sellers and separate doers contradicted the seller-doer philosophy.

As much as I loved my 20-plus years with Unisys, it had become a miserable place to work. It became more and more clear that Richard and I, and many others like us, did not fit in the new seller-doer mold.

Soon, the seller-doers were more than just drive-by helpers. The same ones were showing up on a regular basis. They wanted to meet with our Project Managers, sit in our sales meetings, and meet with our clients. They were always on me for "deep-dive" client reviews. I never told them any more than I had to, and I kept them away from our clients as best I could.

Finally, we were down to one seller-doer that just would not leave. His name was Scott Ess. Like the others, he was new to Unisys, knew little about the company, and less about the State of Florida. He told us that he would be here for a while to "help out" and to provide additional support and leadership.

I felt bad for him and went out of my way to be nice. But no one wanted him here, and he did not want to be here either. He was quite open about that. After a couple of months, he was gone, and we never heard from him again.

But they kept coming and another one was on the way. It was like playing Space Invaders. The next seller-doer lived out of state like all the others but was not as subtle about her mission here as Scott had been. Her name was Jessie Jayne.

It was the same drill but this time the seller-doer was here to stay. At least that was what Unisys intended according to

Jessie.

After I ran out of excuses to keep her out of the Department of Children and Families, I relented and set up an introductory meeting with a few of the DCF executives.

On our short ride to the DCF office, she told me that she would soon be looking for an apartment and that Unisys had agreed that she would commute indefinitely.

She introduced herself to the DCF folks as the new Florida Partner, telling everyone that she lived out of state and would be new to Tallahassee. Someone asked if she had started house hunting yet. To my surprise, she told them no, but that she would be starting that process soon. She even asked them where she should look. This was certainly not what she had just told me. She never moved here, and I'm not sure that she ever rented an apartment either.

After a month or so, Jessie invited everyone in the office to an after-work picnic at Maclay Gardens, billing it as an opportunity to relax and promote camaraderie. It was a nice gesture, but most of us would have preferred to go next door to the Cabot Lodge for a beer and popcorn, as we did after work at least a couple of times every week. We all made an appearance at the picnic.

While all of this was going on, I had started to notice slight tremors in Richard's hands that were particularly noticeable when he was holding a sheet of paper. I mentioned it to Madelin. She told me that I should discuss this with him right away telling me that it could be Parkinson's disease or even a brain tumor.

I went into Richard's office at the end of the next day, as I often did, but this time to bring up a particularly difficult topic. Before I had a chance to say anything, Richard told me that he would soon be going to Mayo for some tests due to the shaking of his hands. We talked about it with both of us minimizing the chances of it being anything serious.

Sadly though, Richard was diagnosed with Parkinson's disease. I'm sure the work-related stress at the office only served to aggravate his condition although Richard never said so. After a brilliant career with Unisys, he retired in late 2005 with much fanfare. We threw the biggest retirement party I have ever attended, at the Golden Eagle Country Club. Richard had friends and colleagues from all over the country there to wish him well.

It has been 16 years now. He has done amazingly well and is enjoying his well-deserved retirement. Richard, George Zimmerman, Bob Abernathy, and I (and sometimes others) continue to have occasional lunch meetings at Whataburger, one of our favorite lunch spots from our days at Unisys. We update each other on our grandkids and talk shop about the old days.

Within six months or so after Richard retired, virtually everyone in the Tallahassee office was gone, including me. The steep decline continued with Unisys never returning to its Tallahassee glory days. None of the seller-doers ever moved to Tallahassee.

Chuck Cliburn

30

EVERYTHING WAS BIG

Jan Cassidy, a good friend, and a previous Unisys colleague from Baton Rouge had called me in November of 2005. After a long and successful career with Unisys, she had resigned a few months earlier to take a position with Affiliated Computer Services (ACS) as Vice President and State Client Executive for Louisiana.

She told me that ACS was planning to add a similar position in Florida and wanted to know if I would be interested. Even though she had only been there for a short while, she loved it. She was the lead ACS executive in Louisiana, responsible for the State relationships, overall growth and strategy, and sales across all ACS government business units.

It sounded perfect. I soon flew to Washington, D.C., to interview with Betsy Justus, a high-level executive and wonderful person who would be my first ACS boss. I was offered, and accepted, the position of Vice President and State Client Executive for Florida.

After 22 great years and one horrible one with Unisys, I resigned and started with ACS in January of 2006 for what would be the final and most successful years of my corporate career. My years at ACS also provided the foundation for me to start my own consulting/lobbying business in 2013. I was 55 when I started with ACS.

ACS was a $6 billion, or so, business process and IT outsourcer that did business in both commercial and public sectors. It was one of the largest tech companies in the world that most people had never heard of.

In those days, if you were making student loan payments, ACS was probably processing them. If you were speaking to a call center agent with a credit card company, bank, or government agency, there was a good chance that you were speaking with someone that worked for ACS. When you went through a toll booth or made your payment at a parking garage of a large airport, you were probably interacting with ACS.

For governments, ACS did everything from running Medicaid programs to red-light camera systems. In most cases, ACS had developed proprietary software for its vast array of outsourcing programs. ACS generally avoided traditional systems integration projects. If ACS could not operate the system after it was completed, we were generally not interested. But there were exceptions.

Everything was big. The company only pursued large contracts. For state governments, if the annual contract value was less than $2 million per year, ACS would usually pass.

I was responsible for managing overall growth and strategy, for leading and coordinating sales across all ACS government business units, and for managing executive-level client relationships in Florida.

I did not report to the business units, which meant some loss of control for them. The business units (Healthcare, Transportation, and Government Solutions) had their own sales organizations that worked under the local guidance of the State Client Executives in the larger states. Under the new model, I would implement and drive Florida strategies in coordination with the business units.

This new strategy was championed by Harvey Braswell, a seasoned ACS senior executive. Harvey was well known in the IT industry, especially in Healthcare. He had previously served as President of ACS's Healthcare business unit before heading up the ACS Public Sector Sales organization and being tasked with implementing the new model. He had been with ACS for many years and was one of its most respected executives. The new model was his baby.

Harvey was somewhat of a crusty old dude with a lot of North Carolina country that never left him, and he was as smart as a tree full of owls. Of all the corporate senior executives I ever knew, he was one of my favorites and one of the best. Sadly, Harvey died a few years ago, may God rest his soul.

After I had been with ACS for a couple of months, Harvey held a national meeting in North Carolina to introduce the State Client Executives to each other, to him, and to the leadership of the business units. Each of us made a presentation on our state, including our business plan and forecast.

I was always well-prepared for such meetings, but for this one I was over-prepared. I considered it to be one of the most important internal presentations I had ever made.

The meeting began, as they always did, with Harvey making some opening remarks followed by everyone introducing themselves. I was among some very impressive people.

Fortunately, I was able to see several of my colleagues make their presentations before it was my turn. None of them, except for my friend, Jan Cassidy, knew their states as well as I knew Florida. Jan knew everything there was to know about Louisiana.

I had been in Tallahassee for more than 15 years, mostly with Unisys, essentially doing the same thing that I would be doing for ACS. For me, it was like changing teams in the middle of a game. All I had to do was change jerseys and keep playing, but with one big difference. At Unisys, Florida was always Richard's state. At ACS, Florida was my state.

When it was my turn, I started my presentation by acknowledging that I probably had too much information and too many PowerPoint slides. I had state organizational charts, lobbyist information, key competitors with names, and all sorts of stuff. Not far into my presentation, Harvey politely asked me to skip ahead to the business outlook – typical for an IT sales meeting.

Even though I had been on board for only a couple of months, I had an excellent forecast that was more than just a wish list. I knew the deals that would be coming up where ACS had a good chance to win and was able to discuss them in detail.

Two of those turned out to be huge wins in my first year – the Florida Turnpike Modernization Project and the Medicaid Reform Choice Counseling project. As they say, timing is everything.

One of the corporate executives at the meeting was Tom Davies. Tom worked at corporate in Washington, D.C., as the Senior Vice President of Marketing and Business Development.

Selling Information Technology

He had an EDS background, as did Harvey, and had also worked for the State of Florida earlier in his career. Although he was relatively new to ACS, he was well-respected and close to Harvey. I did not know him at the time, but he would later become a good friend, a trusted advisor, and my boss for a while.

Tom called me a few days after the meeting to congratulate me on my presentation. He shared that I had impressed Harvey and that he had opined that I would probably be the most successful of the group. What a great and welcome change from my last year at Unisys.

ACS had a large operation in Tallahassee, much larger than I had realized. There were probably 800 or so ACS people in town when I started. Most were operational people working on various ACS outsourcing projects.

I knew that ACS was a stovepiped organization, but I did not realize just how stovepiped it really was. Each of the main three ACS businesses was headed by its own Group President. These groups were independent and had little knowledge of or little interest in what the other groups were doing. I found that some of the project teams in Tallahassee were completely unaware of other projects in town that were led by different ACS groups.

Prior to the new model, everything to do with sales was driven from the top with virtually no state-level strategy or execution. Even though ACS was one of Tallahassee's largest commercial employers, few in Tallahassee knew who we were, not even the mayor. When introducing myself as being with ACS, I would often get the response, "American Cancer Society?"

ACS was not part of the Tallahassee business community.

We did not belong to any key business or civic organizations, we did little to help with high-profile charity organizations, not even the United Way, and we had no one serving on any local boards. Additionally, there was no sense of ACS unity or team spirit in Tallahassee.

These were among the many things that I wanted to fix. The unity issue was reasonably easy, but not so with community presence – that required investment dollars that had to come from the business units. I spent years dragging them along, always kicking and screaming. I would have done it without them, but they had the money.

While ACS had an impressive revenue stream with the State of Florida, it had been over three years since they had signed a new contract. Getting ACS back on the winning track in Florida was my top priority.

Not long after I started, the Agency for Health Care Administration (AHCA) released an ITN for the Medicaid Reform Choice Counseling Project. This project fell into the ACS Healthcare Solutions Group and was right in the ACS wheelhouse. We already held the Medicaid Enrollment Broker contract at AHCA, which was quite similar in function and scope and AHCA liked us, which always helped.

It was my first deal to bring before ACS's formal Sales Review Board (SRB). This group of ACS corporate executives had to approve every deal that the company would pursue. It was more formal than the similar process had been at Unisys.

Most of these top executives were in a conference room in D.C. looking at hard copies of PowerPoint slides while it was being presented by someone in the field over a speakerphone. Good deals were almost always approved, but it certainly was not automatic. There were always a few on the SRB that assumed every deal had something wrong with it if they looked hard enough. These were usually the legal and finance people that salespeople referred to as the order prevention department. Fortunately, this deal sailed right through.

Choice Counseling was a Medicaid-related program initiated by the Florida Legislature designed to provide managed care options to Medicaid beneficiaries. It was quite different from the traditional model, complex to explain to beneficiaries, and something that AHCA had no interest in doing themselves.

Therefore, AHCA was outsourcing the administration of the program including outreach, advertising, public relations, and call center operations. The initial pilot would be conducted in Broward and Duval counties with several smaller counties added later.

To a large degree, this was a call center project, something that ACS excelled at. The only aspect of the project that we could not handle internally was the outreach component.

I recommended to our team that we subcontract the outreach work to minority-owned public relations firms in Jacksonville and south Florida. That way, these firms would be knowledgeable of the local markets.

This would be an effective way to address this important issue and could possibly give us a competitive advantage. Everyone agreed, so I took ownership to do the research and the initial interviews with potential firms. My first call was to Windell Paige, Governor Bush's Director of the Florida Office of Supplier Diversity

Windell was appreciative of the call and our interest in including Florida-based minority firms on our team. He gave me the names of several potential firms. After speaking with all of them, two stood out above the others: one in Jacksonville and one in Miami. Both were African American, women-owned firms. We subcontracted with both.

We submitted an excellent proposal but did not score in first place. Still, we scored well enough to make it to negotiations along with two other bidders. So, that was fine. Large ITN-based procurements like this were almost always won or lost at the negotiation table.

We brought the entire team, including the public relations firms, to Tallahassee to do the initial proposal work. Both had put together exceptional outreach plans, so good, in fact, that we decided to take their CEOs with us to the first negotiation meeting. Their plans included specific radio stations, targeted newspapers, bus routes, and churches that catered to likely Medicaid demographic groups in their respective cities and counties.

We then prepared and rehearsed the first negotiation session. We assigned various ACS team members as the leads for each topic, including the outreach component knowing that this might be handed off to the subcontractors if it went very deep. We asked both to practice responses to likely questions but not to speak in the client meeting unless we called on them.

Within a month or so we were at the first negotiation meeting. We were pleased that the meeting began with more of a casual and friendly atmosphere than they sometimes did. This was probably because they knew us. We were ready and confident that we would quickly make up any ground that we may have lost with the proposal.

After 30 minutes or so, AHCA turned the discussion to our outreach approach. The first question they asked was a giant softball, perfect for our two subcontractors. AHCA told us that their biggest concern was the ability of the selected vendor to reach the right people and to get them into the program since the State could not mandate enrollment. They were quite open about the fact that they liked the approach we had outlined in the proposal.

We immediately handed this off to the subcontractors and they hit it out of the park. Marketing and outreach ended up being the main topic of discussion for the rest of the meeting.

While there were still other significant areas still to be discussed, we left the meeting feeling happy about our chances, bordering, in fact, on downright giddiness.

The remainder of the lengthy negotiation process went well, and we continued to feel better and better about our chances with each step along the way.

Our giddiness was justified. On April 17, 2006, ACS was awarded this $50 million contract. We won!

I was also aware of a toll system project coming up with the Florida Turnpike Enterprise (FTE) but did not have many details. All I knew was that the ACS Transportation Solutions Group (TSG) was a major player in toll systems and that this project was going to be huge – probably $100 million and maybe more. It was the Florida Turnpike Enterprise Modernization Project.

I called Betsy to discuss it and she told me who to call in TSG. I was working on this deal parallel with the Choice Counseling project.

TSG was a global leader in toll system operations. I think it is fair to say that ACS was considered by many to be the industry leader. Many of the large toll systems in the northeast were operated by TSG at the time. Michael Huerta, an impressive executive who had held several top positions in the U.S. Department of Transportation during the Clinton administration, was the Group President.

I called the person Betsy had suggested, and it was a less-than-encouraging discussion. They already knew about the Florida deal and had already decided they were not going to pursue it, even though the bid was not yet out. I was given a list of compelling-sounding reasons to justify their decision.

I was told that the FTE did not like ACS because we had gone over their heads to the Governor a few years earlier to pitch outsourcing, the Florida (Republican) administration did not like Michael Huerta because of his Democratic ties and background in the Clinton administration, and one of the incumbent technology vendors was loved by FTE. Plus, this would be a large SI project and not a BPO project.

I immediately had one of our lobbyists set up a meeting for me with the Secretary of the Florida Department of Transportation (DOT). The Florida Turnpike Enterprise reported to DOT.

I soon met with him and learned that he was quite knowledgeable about ACS. Actually, he seemed to know more about our background in tolling than I did. We had a great open discussion that confirmed to me that we should bid on this project.

I also brought up the incumbent issue, and I will never forget his response. "Chuck, that would be like me wanting Sears to build my house because I liked their garage door openers."

He certainly indicated no preference for a particular vendor, he knew who we were, and he was aware and complimentary of the work we had done in other states. He assured me that there were no favorites.

I called Betsy to tell her about the meeting with the Secretary. She conferenced in Harvey, which was my first call with him.

I went through the whole thing, explained that TSG had already decided to no-bid, the reasons why, and what the Florida DOT Secretary had told me.

Harvey was a colorful man, and this was a colorful conversation. "Who decided to no-bid?" he asked.

"TSG," I replied.

"No, WHO decided to no-bid?"

Harvey told me to send in the request to get this opportunity scheduled for the next SRB "preview" and to start working on the deck. The "preview" was the first step in the deal-review process designed to be held prior to the bid being released. Unlike Choice Counseling, this one would not be smooth sailing.

Harvey told me he would give Michael Huerta a heads up and, with some obvious agitation, told me that only the SRB could make bid/no-bid decisions. I had a bad feeling that my first TSG deal review was going to be Harvey and me against the entire TSG organization, but at least there was a chance to get this turned around.

It would also be one of the early test cases of the new sales model. Under the previous model, this deal would have never seen the light of day in my opinion.

I scheduled the presentation for the next SRB. Harvey and the other SRB executives were on the call, including Michael Huerta and a few members of his leadership team.

I made it to slide two or three before the meeting came off the rails. From there on, I was mainly an observer. There was one camp, led by Harvey, that felt we should bid on this, and another camp (TSG) that felt we should not. It was a great exercise in observing the ACS power structure in action.

It was clear that Harvey was not going to relent. After quite a long debate, someone in TSG suggested that, as a compromise, we try to bid as a subcontractor with the idea of doing all the "conventional" toll work. The TSG guys felt that we should partner with a large federal contractor that would be more interested in the open-road tolling work and not the conventional tolling stuff where we excelled. They had a specific contractor in mind. By the way, "conventional" tolling basically meant any toll booth you could physically see. I had to check on that after the meeting.

TSG took the action item to reach out to the potential prime contractor. I was skeptical, suspecting that they would not be interested and that this was mainly a clever ploy for TSG to get out of the deal.

But to my surprise, the potential prime contractor was interested and so was TSG. Negotiations got underway to develop the teaming agreement and the scope of work. The ACS content of the deal was roughly $100 million. Working together, the two companies put together an impressive proposal.

I was there with our team for all the pre-bid meetings, the vendor presentations, the bid opening, and the final vendor selection meeting. All the meetings were held at the FTE administrative headquarters in Ocoee, Florida, near Orlando.

It was not typical for ACS to be a subcontractor, and it made us all uncomfortable to see another firm carrying the ball. ACS, in my opinion, did a better job of presenting our content than did the prime and it looked to me like FTE felt the same way.

During the final public meeting, there was an open discussion among the evaluators before the decision was announced, and it seemed likely that we were going to win. They huddled and then announced the decision. We won!

In October of 2006, our prime, with ACS as a major subcontractor, was awarded a 10-year $200-million-plus contract to modernize the Florida Turnpike system.

Elated, I was soon in my car calling Harvey, Betsy, and Tom to give them the good news. I wanted to tell them before they heard it from TSG.

In the coming months, I drove to Ocoee periodically to meet with one of the key FTE officials that I met during the sales process. He was always cordial and expressed their happiness with their decision but repeatedly expressed his surprise and disappointment that ACS had not bid as the prime. The Florida Turnpike Enterprise became a good customer for ACS for years to come.

Chuck Cliburn

31

SOMETHING GOOD FOR FLORIDA

Not long after these two wins, Harvey was in Tallahassee and wanted to meet with me while he was in town. He told me that he had no formal agenda and that he just wanted to "catch up." This would be my first face-to-face meeting with Harvey since the two big wins, so I was looking forward to it.

However, Harvey did have an agenda and the meeting set a trajectory for me that I would work on for several years.

We soon met in my office. In Harvey's plainspoken way, he told me that ACS had all sorts of corporate facilities around the country and including some states that did little, if any, business with ACS. As he put it, "Some of these states don't buy shit from us." He said that Florida had always been good to us and that it was time we did something good for Florida.

All our facilities in Tallahassee were tied to projects. In general, when the project went away, the facility and staff went away. Harvey wanted to open something here that would make a statement, would not be tied to Florida projects and would show a corporate commitment to the State of Florida.

He told me to start working on a Florida-based Center of Excellence (COE), an upscale facility that would serve as a showplace production site and visitation center for ACS prospects and clients from across the country. He said it had to be something big and something impressive.

He cautioned me against letting the business groups or the ACS Facilities Group try to force this into a dumpy site like we used for most of our operational projects. He assured me that the rest of the top leadership would be onboard.

That was the extent of his direction, but it was all I needed. It was an exciting and challenging business project and something that I really wanted to do.

When he left, I called Betsy, Tom, and Madelin. Betsy tried to act surprised and told me that Harvey liked me and to get on it. Tom was enthusiastic and likewise tried to act surprised. He quickly became a strong supporter of the project and was very helpful all along the way. I think Madelin's response was, "That's cool."

The COE idea fit in perfectly with my plans and the work I had already done to elevate the ACS image and team spirit in Tallahassee.

I had established an ACS Leadership Board with representation from all the local projects. We met periodically so that everyone could share what they were doing. I always provided an update on the deals we were working on and on other ACS issues that were of common interest to everyone. The status and progress of the Center of Excellence (COE) became a favorite topic, and sometimes the only topic.

At my recommendation, we had already joined Florida TaxWatch and Associated Industries of Florida, two of Florida's most influential organizations. Both had substantial sway with state government. I would eventually sit on both boards.

Soon, we became a founding member of the Tallahassee Minority Chamber of Commerce where I was also on the Board of Directors. We became a participating firm and significant fundraiser with the United Way. And for the first time, ACS became involved with the local chapter of the American Heart Association where I served as the Chairman of the Tallahassee Heart Walk in 2009.

Florida TaxWatch named me to their Government Cost Savings Task Force. This Task Force consisted of top executives from leading corporations across Florida as well as high-level state officials. The group submitted a yearly report of recommended cost-savings measures to the Florida Legislature for their consideration.

We were earning our way onto the Tallahassee map in a big way. People knew who we were, especially government and civic leaders. But now, the COE had added rocket fuel that could put ACS into a league of its own in Tallahassee and with the government of the State of Florida.

While ACS was a big company, we did not have a big company culture. There was an entrepreneurial spirit mostly absent of power grabs and fiefdoms that can often impede progress in large companies.

I worked closely with the Mayor's Office and with the Tallahassee/Leon County Economic Development Council. There was the potential of hundreds of new jobs tied to this, maybe even more over time, so it was a very big deal in Tallahassee.

Both organizations became ambassadors for us. The mayor of Tallahassee, John Marks, became involved. I addressed the City Council on multiple occasions regarding ACS, our presence, and our plans and progress. I was a guest on Florida Public Radio for an interview segment regarding ACS and our Florida plans.

I did as much as I could, as quickly as I could and as publicly as I could, in part, to make it as difficult as possible for ACS to get cold feet and back out.

Our commercial real estate firm did some initial research for me and took me to look at several potential sites. All were terrible. One of them was across the street from an adult bookstore. We wanted something that would be at least 40,000 square feet and something impressive.

I called Rick Kearney, the owner of Summit East Technology Park, the most upscale office park in Tallahassee, in my opinion. I didn't know Rick at the time, but I was well aware of who he was and his background in IT and in commercial real estate. In addition to owning Summit East and other commercial properties, Rick was the founder and chairman of Mainline Information Systems, one of the world's largest IBM partners.

He showed me one of their new buildings that had over 40,000 square feet of contiguous space that was not yet built out – it looked like an indoor football field. Rick told me that they would build it out to our specifications and amortize the build-out costs into a long-term lease if we wanted to structure it that way. I knew ACS would like any option that minimized cash outlay.

After Rick's tour and overview, I knew that was the place. It became my mission to get ACS into Summit East. I spent a lot of time on this project, but it was always secondary to my responsibility of growing our business with the State of Florida.

Chuck Cliburn

32

"THEY WENT UP"

For the next seven years, we dominated the State of Florida marketplace for large multi-year BPO and IT contracts. It was an extraordinarily successful time for ACS and me. One of the largest and most unusual wins was at the Florida Healthy Kids Corporation (FHK).

FHK was created by the Florida Legislature to administer the Children's Health Insurance Program (CHIP). It reported to a board chaired by the Chief Financial Officer of the State of Florida.

The final selection and award process for this one was probably the most unusual of my career.

FHK outsourced most of the administration and operations of the corporation. In late 2006 they released an ITN for a new contract. The new BPO contract would replace the contract that was in place with the incumbent vendor.

By all logic, this project was a natural for us. It had a large call center component, other similarities to projects ACS operated in Florida, and we had CHIP experience in other states. We felt we had an excellent chance to win this project.

But, as always, we had strong competition. While there were not many firms that pursued CHIP deals, the ones that did were very good.

We knew, or thought we did, that FHK had not been particularly happy with the incumbent vendor. This turned out not to be true, at least among the FHK evaluation and operations team.

Soon after the proposals were submitted, we learned where we stood from the public meetings of the evaluation committee, and it was not good. To make things worse, FHK had advised all three vendors of the pricing but without vendor names. We were roughly $10 million higher than the lowest bid, which we correctly assumed was the incumbent. Our price was roughly $95 million. The third vendor was somewhere close to us.

It was easy to monitor our progress and the direction of the evaluation from the public meetings. For the public evaluation meetings, FHK had set up a makeshift screen separating the evaluators from everyone else so the vendors could not see them, but we could hear everything they said. They designated each vendor with the name of a pro football team so we would not know "who was who." That, of course, did not work and it took about five minutes to identify which vendors the football teams represented.

The outlook was getting worse for us by the day. Our only hope was for some sort of divine intervention. Maybe the incumbent would withdraw. Maybe our BAFO (best and final offer) would be good enough to change their minds. But something big would have to happen. Any real chance of winning certainly appeared to be a longshot.

By the time the negotiations and decision day came, we were as prepared as we could be to lose. Still, our Corporate Chief Operations Officer was on standby for us to call with updates during breaks and so he could provide any last-minute approvals if needed.

The decision day was a day of uncharted waters. Since FHK was a corporation, all it would normally take for the final decision would be a motion and a vote by the board. But that is not what happened, at least not for the first four or five hours.

The Florida Chief Financial Officer (CFO) is an elected cabinet position in Florida. By statute, the CFO also serves as the Chairman of Florida Healthy Kids. I had never seen the CFO at any of the previous procurement or board meetings, but he was at this one.

The board members took their seats at a table facing the vendors with the CFO seated in middle. He chaired the meeting and took full control.

After calling the meeting to order, he got right to work by individually asking each board member this simple and direct question: "Are all three vendors qualified and capable to deliver this project?" One by one, they all said yes. He then said something like, "Good. Now we are down to contract terms and price." It now seemed obvious that this was not going to be a slam-dunk for the incumbent after all. Maybe we still had a chance.

He then began the long and laborious line-by-line contract review, which took a couple of hours. In the end, all three vendors agreed to all the contract terms. Now, they were apparently down to price.

After some discussion, the chairman announced a 30-minute break for the vendors to make any desired last-minute changes to the best and final prices we had brought with us. He followed by saying that the vendor with the lowest price would be awarded the contract. We now had an equal chance to win.

We found a secluded area and called our COO knowing that the price to beat was $85 million based on the pricing they had previously provided to the vendors. Our COO reasoned that the incumbent would probably lower their price by another million dollars just to be safe, so he instructed us to submit a price of $83 million. This was obviously a huge drop in our price and one that nobody expected. Including me.

We returned to the meeting room with our new $83 million price handwritten on a piece of paper. The meeting was called back to order.

To the surprise of everyone, the third vendor withdrew from the negotiations. We surmised that their corporate folks had not liked some of the contract terms that their team had agreed to during the earlier session. Our company and the incumbent, the remaining two bidders, were asked to bring forward our final pricing.

The room was extremely quiet as the board members gathered around the CFO to look at the two sheets of paper with the final prices. The tension in the room was unbelievable. I felt like I was going to pass out as I am sure everybody did.

Then, we heard one of the board members whisper, "They went up." We didn't go up. We knew we had won if the CFO held to what he had said.

The CFO announced that ACS had the lowest price and asked for a motion to award the contract. Going against what the CFO had said, one of the board members commented that there were things in play other than price and made a motion to award to the incumbent based on best value.

Now, the tension was even higher. It was like nothing I had ever experienced. Before anyone could second the motion, the CFO made a statement along the lines that every dollar spent on a vendor was a dollar less for their real mission. This may not have been exactly what he said, but it was clear to most of us that he did not favor awarding to a vendor with the higher price. There was no second, the motion to award to the incumbent was withdrawn, and a new motion was made to award to ACS. It was seconded and approved. We won!

Never I had come so close to losing then still won. I'm sure the opposite was true for the incumbent vendor. I felt bad for them but soon got over it.

On the way back to the office with my colleagues, I called Betsy to tell her. Like the rest of us, she had no real expectation of winning. At first, I do not think she even believed me. She was surprised and very happy.

I guess things have a way of balancing out. This made up for some of the deals throughout my career that I had lost after having been certain that I would win.

We had now won three consecutive large deals, and we continued to win over the coming years. Large competitive deals still to come included the Unemployment Insurance Debit Card payment program, the Third-Party Liability project, the First Notice of Loss Call Center, and the Statewide Email Outsourcing Project. We won them all and some others too.

We were also awarded a large multi-year contract renewal with one of the State's largest agencies after all option years had been expended. It took some great work by our lobbyists and legislative action to make that happen.

All this success helped propel my efforts with the Center of Excellence. We were on a roll and Florida had become one of ACS's top states. Few, if any, vendors rivaled our success in Florida state government.

33

CHANGE, THE ONLY CONSTANT

Somewhere in the 2008 timeframe, I got a call from Betsy and immediately knew that something was wrong. She was always cheerful and could brighten your day with just the tone of her kind and gentle southern voice. But this day was different. She was crying.

She told me that she no longer worked for ACS, as she had been fired just a few minutes earlier. I was shocked and didn't know what to say, other than I was sorry and to ask what had happened. She told me who had called her and that she was, in effect, told that she was no longer needed. Betsy loved her job and the company, and everybody loved her as far as I knew.

I talked with almost everyone I knew about it, but I never learned the real reason for her departure. I can't imagine that it was for a good reason. Whatever the reason was, it was a big loss for the company.

Shortly before this happened, she had promoted me to Senior Vice President.

Not long after this, Harvey retired and bought an IT firm in North Carolina. Betsy soon joined him as Vice President of Sales. A few years later she died of pancreatic cancer. May God bless her soul.

Joe Doherty had joined the company as the new Group President coming from another large IT firm. Tom Davies worked closely with him and was, in my view, his go-to person, at least for a while. Tom was also the Corporate Executive Sponsor for Florida and a strong advocate for all I was doing, including the COE project.

I got to know Joe quite well and I liked him. By this time, we had made significant progress on the COE and Joe supported the project, due in part, to Tom Davies, I'm sure.

I was soon working with David Nice, the head of ACS Corporate Real Estate and Facilities on the COE project. David was the top facilities executive in the company and was one of the nicest and most helpful people that I ever worked with.

When he called me to say that he wanted to come to Tallahassee to tour the city and the properties on my short-list, I knew we were close to making this happen. I arranged a half-day agenda with the help of the Mayor's Office and the Economic Development Council.

Our day started at City Hall with lunch and a presentation by the City and the EDC. This was followed by a guided tour of the city in a small city bus with visits to the top three potential sites. I saved my pick, Summit East, for the last stop.

Rick Kearney and his Summit East executive team gave us the tour and made a brief presentation. I had done a lot of selling to David before we got there. As soon as the elevator opened to the 40,000 square feet of new and open space, David said something like, "This is what we need." No other property was under serious consideration from there on.

It took months, but ACS eventually signed a lease agreement with Summit East. A project management team was assigned by the ACS Facilities Group and things really started to happen.

ACS had several similar COE sites around the country which served as somewhat of the starting point for the design of the Tallahassee facility. The facility people knew what they were doing and never tried to cut corners for the sake of going cheap.

There were several people on the executive leadership team that had a big stake and a big say in the project, but none of them had much interest in being personally involved. Plus, I was in Tallahassee and none of them were. I probably had more control than Harvey had even envisioned. With Harvey gone, I was the driving force behind the project. Except for the facilities group, everyone else was following along and getting involved only when they had to. That was a good characteristic of ACS. There was not a lot of bureaucracy to get in the way.

I remember being asked one day to meet the ACS architect at the facility, which was still nothing more than a giant empty space. I met him in the lobby, and we took the elevator to the third floor, where the COE would be located.

When the elevator door opened, he held it open and asked me what I envisioned from that vantage point. Fortunately, I had already thought about that. I told him double glass doors directly across a hallway opening to a reception center with a direct view to the receptionist desk on the opposite wall featuring a large and impressive ACS Center of Excellence logo above it.

He was taking notes on a legal pad and asked me what else I envisioned. I suggested two executive offices and a large conference room in the reception center. He then told me about some of the other things they would be including such as a nice locker room for all call center employees. He then walked me around the large empty space and described his vision of the plans for the rest of the facility.

It was not long until we met again at the job site to review the initial building plan and it was awesome. The reception center design was perfect, and the two offices were bigger than I even hoped for. The other management offices would be in the secured area, near the production teams.

It took another year or so for the construction to be completed and for us to begin to transfer operations to the COE.

ACS had operated many of the call centers for our government debit card programs in Texas but would be transferring some of these programs to the COE. There were other operations transferred to our COE as well, including an emergency call center for a power company in the mid-Atlantic area.

We proposed the COE as the operational center for Florida's debit card program for Florida's Unemployment Compensation program, and for the Florida Enterprise Email Project that we were working on at the time. Harvey's vision of "doing something good for Florida" was paying off. We won both! Additionally, the COE was designated to be the operational center for future email projects with other state and local governments.

While the construction was in process, I received a call from one of the executives in the Transportation Solutions Group. He told me that they were planning to relocate their Maryland-based airport parking test lab and development facility to a new site and that they had heard about the Florida COE. He was interested in learning more about it as a possibility for them.

I described the facility to him and told him how much-unallocated space we still had remaining. He was very interested and soon came to Tallahassee for a closer look.

I coordinated the TSG visit with Summit East and made sure that they got red carpet treatment as David Nice had received. It was a complete success and TSG soon decided to relocate to the Summit East Office Park.

However, they felt like the COE may not have enough space for their future plans, so they chose a separate space in an adjacent building. TSG was very independent, so they probably preferred to have their own building anyway. But regardless of reason, that was even better from my perspective as it resulted in an even larger ACS footprint in Tallahassee. TSG soon moved into their new Summit East facility.

We were on track to easily become Tallahassee's largest commercial employer. I envisioned Summit East as someday becoming one of ACS's largest corporate facilities. It could have not been going any better.

Chuck Cliburn

34

THE BEGINNING OF THE END OF ACS

On the morning of September 28, 2009, I woke up early, grabbed the remote control, and turned on the bedroom TV to CNBC. It was my ritual.

Somewhere between awake and asleep, I heard this: "Xerox Corporation announced today that it is acquiring Affiliated Computer Services, a Dallas-based business process outsourcing firm." I was hoping it was a nightmare. I should have been careful of what I was hoping for. It was the first day of the beginning of the end for ACS.

I got up, got dressed, and went straight to the office. I called everybody I knew that might know something about this, including Tom Davies and Joe Doherty. Tom quickly called me back and did his best to put lipstick on the "situation." I was not happy. Things were going great for ACS in Florida and I did not need a merger with Xerox, which I perceived as a copying machine company, to muck things up – strictly my view, of course.

ACS and Xerox had a well-scripted message for ACS employees. This will be great, we will continue to operate on a business-as-usual basis, we will keep our brand, and so forth. I didn't believe any of it.

For a while, some of this was true. We kept our logo and operated as ACS, a Xerox Company. But predictably, the ACS name and logo were dropped after a while.

The cultures of these companies could not have been more different. ACS was not a process-driven company. ACS empowered people to do their jobs and to make things happen.

Xerox was quite different. It had rules, policies, and procedures for everything. It seemed to me that it was better to follow the rules and do nothing than do something that might break "a rule."

As an example, there were lots of rules about who would have an office and who would have a cubicle, how big they would be, and so forth. Not long after the merger, someone from ACS facilities called me to tell me that my new office in the COE was far larger than what was allowed by Xerox policy. He said it was just a heads up. It was low on my list of things to worry about.

The local Xerox people told me about their governmental affairs group, the group that hired, fired, and managed their lobbyists. Xerox had a much different view of how lobbyists should operate than ACS did. There were lots of lobbyist rules, and I didn't like any of them.

They told me that the Xerox sales organization could not engage with lobbyists without going through the governmental affairs staff first and that they could not direct lobbyists to do things. It all seemed so nonsensical to me that I just assumed they were exaggerating.

ACS had several lobbying firms that I worked with on a

daily basis. I treated them as members of my team, and they did the same.

But the Xerox salespeople were right. One of the first corporate casualties of the merger was the ACS governmental affairs group. Xerox governmental affairs was soon in charge.

Maybe six months or so after the merger, there was a company-wide call for the ACS sales organizations to be led by Xerox governmental affairs. This was the meeting where we would all be given our new marching rules for working with lobbyists and our government clients.

It seemed borderline absurd to me, and I was not by myself. Governmental affairs would manage the activities of the lobbyists, not the people in the field. Salespeople, no matter how high in the organization, could no longer meet with elected officials unless it was approved by governmental affairs. We would be accompanied by a lobbyist for such meetings unless there was approval to the contrary. It seemed to me that Xerox did not understand how to succeed in the governmental BPO world. One might easily conclude that I was right.

Some eight years later, Xerox returned to its copy, print, and document roots and spun off what was once ACS.

It did not take long for me to start getting calls from Xerox about the COE. At first, it was requests for the history of the project and copies of the contracts. I got several calls asking me if we had considered the XYZ facility because it would have been far more "cost-effective." There was some "second-guessing" going on.

Once, they called me to tell me about a facility that was becoming available that they pitched as sounding "ideal." It was one of the less-than-impressive facilities that ACS was vacating to move into the COE. I was not impressed with their real estate and business acumen.

Eventually, they quit calling me about the COE, but they were not fans. To them, it seemed to be just an expense and a decision they did not make, not the strategic move that Harvey Braswell had envisioned.

But by now we were moved in, and it was a beautiful facility indeed.

For a year or so, we (ACS) had been planning a ribbon-cutting ceremony with lots of media coverage and remarks from the Mayor and probably the Governor or his designee. Had Xerox allowed it, it would have been carried by Tallahassee TV stations for sure and probably other Florida stations as well. It would have had much print coverage as well. The Governor's Office and the Agency for Workforce Innovation would have likely been all over it, but Xerox was opposed to it, so it never happened.

In my opinion, that was a huge mistake. If Xerox had engaged the Governor and other state officials in the opening of the COE, the upcoming cancellation of the Florida email project may well have been avoided.

The people that made ACS "ACS" began to leave. The ACS entrepreneurial spirit was slipping away. The ACS I loved was gone.

35

THE FINAL FRONTIER - FLORIDA ENTERPRISE EMAIL

In 2010, or so, we began to work on a project that would consolidate and outsource the operations of the State of Florida's email systems. It would be the last big project of my career.

As with most Florida systems, each agency operated its own email system. The Florida Legislature viewed this as an opportunity to reduce overall operational costs and to modernize and consolidate the email system.

In somewhat of an unusual move, the Legislature passed legislation requiring this project to be done with certain legislative checkpoints along the way and with final legislative contract approval. Once the final award was made, the Agency for Enterprise Information Technology (AEIT) would make a final report to the Legislature of the estimated cost savings for their approval and funding. Until that step was completed, there could be no contract.

Most of the procurement work was delegated to the AEIT with the help of an evaluation panel of state officials from several agencies and other entities. The final contract and operations of the project were delegated to the Southwood Shared Resource Center (SSRC), one of the two-state data centers at the time.

After a year or so of prework, the ITN was released. The State cut no corners and was envisioning and requiring a world-class email system that was far more sophisticated and feature-rich than any of the State's existing email systems.

As multi-agency projects sometimes do, the project quickly became controversial. Some agencies had little interest in giving up their internal systems. They inherently did not like giving up control of anything, even email.

This was just before cloud-based email systems were widely accepted, so Florida indicated a preference for a private-cloud approach. They wanted a private data center that would not be shared with other entities and a redundant backup data center as well.

The State required that costs of securing and outfitting the new data centers and other related start-up costs be included in the per mailbox/per month price. As a result, the project required an unusually high upfront capital expenditure that would take years for the winning vendor to recover. Only the "big boys" could afford to bid on the project.

We were well-qualified in email outsourcing, but particularly in the commercial sector. In fact, Disney was one of our references, which never hurt in Florida.

Like most, it was a long and grueling procurement process, but when it was finally over... we won!

This was another very large contract. We estimated that it would exceed $100 million by the end of the contract. There was not an exact number since the ultimate value would be determined by the number of users and the options they chose. Additionally, the contract was also made available to Florida cities and counties, potentially making the contract value even higher.

For reasons that I never fully understood, there was obvious animosity between Dave Taylor, the State CIO (and Director of the AEIT), and the Legislature. Having known and liked the people on both sides, it always seemed to me that there was more animus toward Dave than there was the other way around. But for sure, some of the House people just did not like him.

The Legislature's main role in this was to approve the cost savings analysis that was completed by the AEIT and to approve the final funding.

As sometimes happens in government, the "cost savings" soon became the only real justification for doing the project at all – or so it seemed. Somehow, the fact that Florida's email systems were siloed, old, and inefficient did not seem to matter much anymore.

I often told my state friends that sometimes you must replace your roof even if it costs more than keeping the one that leaks.

It was this obsession with cost savings and the poor job that was done with the State's internal cost justification that ultimately doomed the project. There were several forces at work that eliminated any reasonable chance of an objective cost-savings analysis.

First, the only cost savings that were supposed to matter were the total statewide costs. This approach had always been problematic in Florida due to the absence of statewide accounting and budgeting for such things. No one had that number.

Therefore, the data that was used to compare the old costs with the proposed new costs was obtained on the honor system. The AEIT sent spreadsheets to the agencies for them to document the costs of their existing systems including staffing costs, hardware costs, software costs, and so forth. It was basically the "fill in the blanks and we'll trust you" method.

Second, there were few, if any, cost adjustments required for agencies that had "leaking roofs." Obviously, any cost comparison of a new roof to a "leaking roof" is not "apples to apples" unless it is acceptable for the "old roof" to keep on leaking.

At a minimum, the cost of any roof repairs that would be needed over the course of the 7-year contract should have been added to the existing costs. You get the idea.

And finally, once all this data was compiled by the AEIT, the Legislature was not going to believe anything coming from Dave Taylor anyway. The State should have hired an independent auditing firm to do the cost-savings analysis, but they chose not to.

The final cost data showed small cost savings, but the Legislature, especially the House, was not buying it. Some were quite vocal in their assertion that the AEIT had somehow cooked the books to make the costs of the new system look more favorable. Arguably, it was probably the other way around. The cost of staying with the old systems for seven more years was probably more than it appeared.

Some members of the legislative staff publicly opined that the cost of the winning proposal (ours) was too high, citing various publications that indicated lower costs for cloud-based email systems – specifically Google mail. But, of course,

Google did not bid and if they had they would have been disqualified for not meeting the requirements of the bid. We had the lowest costs of the proposing vendors as I recall.

All of this unfolded in a very public way and we feared that the Legislature would not accept the AEIT cost evaluation and not fund the project. The Legislative budget folks seemed adamantly opposed to it, as did some of the agencies.

After the award but before the funding decision and contract, I met with some of the most vocal agency CIOs that were opposed to the new system. The argument was always the same. Their costs would go up and their existing email system was just fine.

They all acknowledged that they did not have a fail-over email data center, or virtually unlimited retention of historical emails, or a periodic built-in hardware refresh – all things they would have with the new system. But that was an inconvenient truth that just did not seem to matter.

The average monthly cost per mailbox for the new system was somewhere around $8. According to the honor system numbers, some of the agencies had higher internal costs, so they were fine with the new system. Others though claimed costs as little as $2, so they wanted nothing to do with the new system. Few agencies cared about "statewide savings." They only cared about the costs for their agency.

Other than the Legislature, the State had no one with the authority to make this happen, as is often the case with Florida enterprise projects. It would be hard to imagine General Motors, Walmart, or any Fortune 500 company operating this way.

The final approval (or disapproval) for the contract and

the funding would be made by the Legislative Budget Commission (LBC) in the fall of 2011. The LBC is a joint committee of the Florida Legislature that has the authority to convene during the year and act on certain legislative issues. For this particular year, the Florida Enterprise Email issue was one of them.

As an understatement, the Chair of the LBC was not a fan of Dave's and hence, in my opinion, not a fan of the project. Without the vote of the Chair, the project was doomed.

We had all our lobbyists working on this, and it was clear to some of us that the project was in big trouble.

On the day of the LBC meeting, I met two of our lobbyists at their office to walk across the street to the Capitol for the meeting. As we were walking, I asked them for their predictions on the meeting outcome. One of them replied with what sounded like harp music from Heaven.

Maybe he was clairvoyant or just feeling lucky, but he said that he knew exactly what was going the happen. Dave Taylor would make his presentation. Then the Committee Chair would publicly embarrass him before reluctantly stating support for the project. After warning him not to be joking around, he assured me that he was not kidding.

That was exactly what happened. Dave Taylor did a good job, in my opinion, presenting the business case to the committee. But when he finished, the Chair lit into him like a misbehaving school kid. Everything Dave had said was called into question. It was not pretty. After the rant, the Chair reluctantly stated support of the issue. A favorable motion was made and approved by the committee. Finally, we crossed the finish line! At least, we thought we did.

We soon had the contract and completed the section of

the COE that would be the primary operational site for the project. We brought in the Project Manager (PM) from the Disney account, finalized all the subcontractor agreements, and after a few months, we were transitioning state employees to the new email system.

Everything was going fine but with the usual bumps and glitches of a large project. The most significant problem was the resistance that our project team was getting from some of the agencies that were never on board with the project in the first place.

Our PM was now responsible for the project, but I was helping her with the State relationships and politics. Coming from Disney, she had no knowledge of the history of the project, nor did she know anyone with the State. She often came into my office to vent her frustration telling me that she had never been involved in a project where the client did not want to implement the system they had just signed up for. Neither had I.

We soon started to hear rumors that the Legislature may introduce a bill to abolish the AEIT and possibly defund the email project during the next legislative session that was now only a few months away. The state agency that held the contract, the Southwood Shared Resource Center (SSRC) would not confirm or deny anything.

One of our best lobbyists agreed to meet with one of the legislators that seemed to be opposed to the AEIT, and possibly the email project, to see what was going on. They soon met.

According to our lobbyist, this legislator told him that their main issues were not with the project or with Xerox but were with Dave Taylor. He was their target. Our lobbyist then asked if the actions they planned to take could be done without hurting Xerox and the project. The response, according to our lobbyist, was yes. We interpreted this to mean that Dave Taylor, and even the AEIT, may be in trouble but that the project would not be canceled.

But we were wrong. We were thrown under the bus with Dave Taylor.

The Legislature introduced a bill to abolish the AEIT and create yet another new IT agency. The bill passed but was vetoed by the governor. The net effect was that the AEIT survived in name only and the email project would soon be dead. The AEIT was later abolished by legislation.

Technically, our contract was still in effect as the State had not yet officially canceled the project. We assumed that the State would use non-appropriation of funds as the basis to cancel the contract and, to my knowledge, that was ultimately the legal basis for the contract cancellation.

The Governor's Office had avoided engaging with us until it was too late, for all practical purposes. But finally, they assigned a high-level and respected state official to work with us in a last-minute attempt at contract revisions that might be more acceptable to the Legislature.

As main points of concession, they wanted a shorter term for the contract (five years instead of seven) and a lower price, arguing that these concessions would make for "better optics" with the Legislature. However, "better optics" for them would have been "optics" for us that were way out of line with our winning proposal and contract.

While this was still going on, we received an official and very public notice of contract termination from the Office of the Governor. Apparently, the right hand did not know what the left hand was doing. Or, perhaps, the right hand did not like what the left hand was doing. In either case, that was the end.

During those final months, our government affairs group had retained one of the best-known lobbying firms in Florida to help, without my input or knowledge. A Xerox corporate executive was sent to Tallahassee to help "save the project," also without my knowledge or input. Overnight, I was excluded from all company calls and meetings on this topic. It was clear that someone was going to take the fall for this, and it was clear who it was going to be.

This was more than just a canceled contract. It was a financial loss for Xerox. We had invested heavily in start-up costs that would not be recouped until years into the contract. That was ok, we signed up for that going in. But it was not ok, in my view, for this to unfold the way it did.

Soon, I was out of the picture but still with the company for a while.

One morning in November of 2012, seven months after the contract was canceled, my boss called me on my cell phone just as I was unlocking my office door. He told me to hang up, close my door, and call him from my office phone.

He told me that I was included in the company layoff that was happening that day. He was apologetic and assured me that it was not a performance issue. He said the company could no longer justify my Senior VP position dedicated to Florida based on what had happened with the email project. Technically, I was "laid off" as a result of the corporate downsizing. But practically, I was fired in a way that made it almost impossible for me to sue Xerox.

I was not surprised, but I was angry – not about being let go, but about how Xerox had handled the entire email situation.

Honestly, Xerox had every reason to be bitter. In the view of many, what the State of Florida had done had little, if anything, to do with Xerox, the email project, or with me. Still, Xerox did not handle it well in my opinion.

That was the end of the COE as well. With me out of the way, the COE was gone before it really got started. Regrettably, the State made all its decisions regarding the email project without knowing what Xerox and the COE could have done for Florida, and especially, Tallahassee.

My career in the IT industry was over and I had enjoyed every day of it. It had been more than 38 years since my first day with Burroughs and it seemed like yesterday.

It was truly a blessing to have a career that I loved so much and to have never once dreaded going to work. I remembered every single win and every loss but most of all I remembered my friends over those 38 years who were colleagues, competitors, and clients. Many of them are still close friends.

I gathered all my stuff from the office and went home knowing that, with God's help, my best days were still ahead. I was 62.

36

NEW CAPITOL IT, MY DAYS AS A LOBBYIST

For years, I had planned to someday retire from the corporate world and become an independent consultant and lobbyist specializing in the IT industry. The Xerox experience just pushed up my timeline by a few years.

My vision was to offer business development consulting services, not just traditional lobbying. The reality is that many, if not most, IT firms hire state lobbyists for one main reason – to help them sell more stuff. I knew I would offer a strong and unique background related to business development.

During my many years in Tallahassee, I worked with some of the best lobbyists in Florida. Some were previous legislators or legislative staffers while others had held high-level positions in the executive branch. But they were not business development or IT experts, nor were they expected to be.

I was ready to open my own consulting/lobbying firm

even though I had never worked a single day for government, had never held a fundraiser, and would never be considered by anyone to be a political "insider." I could never compete with these lobbyists on their field, so I created my own.

My value proposition would be to specialize in the IT industry and to offer business development consulting as the cornerstone of my lobbying service. I would help my clients navigate the complex political Florida landscape, the complex procurement processes, and assist with access to the appropriate governmental officials, especially in the Executive Branch. Most significantly, I would assist my clients on pursuit strategies, including partnering and solutions, something that most lobbying firms do not do.

I knew there would be plenty of IT firms in need of this type of service.

I spent most of December 2012 working on logos, brochures, business cards, and my website. I also developed my business plan and leased an office downtown one block from the Capitol.

On January 18, 2013, I registered my new company, New Capitol IT, LLC, with the Florida Division of Corporations and went to work. For those not in Tallahassee, the "new capitol" was opened in 1977. It sits behind the "old capitol" that is now a museum.

Within a few weeks, I signed my first client, ISC, a Tallahassee-based IT firm. The owners, Edwin Lott and Mark Alexander, were friends and previous business colleagues. Their firm had done subcontracting work for me when I was at ACS/Xerox. I will always be grateful to them for stepping up and becoming my first client.

I soon signed another client, and then another, and within

a few months, I had all the clients I could handle. Over the past eight years, I have assisted some of the industry's largest firms with their efforts in Florida and smaller firms as well.

My consulting/lobbying work has been rewarding in every way, even exceeding my highest expectations. It has been the perfect way for me to utilize my experience and to enjoy my final working years as my own boss. And I am a very good boss.

Essentially, I sell slices of my time doing what I did for Unisys and ACS (Xerox) with some additional legislative work. The difference is that I am now a mile wide and an inch deep instead of the other way around. I work on different issues for multiple clients at the same time. My job is to help my clients win and succeed in Florida.

I am often asked when I will retire. My standard answer is: "Whenever I get tired of what I am doing or when I am no longer useful, whichever comes first."

But when that day comes, I will not stop selling. After all, selling is simply the art of persuading others to see things from our point of view. At the end of the day, we are all salespeople. Some of us were just blessed to get paid for it.

I wish you all happy selling. I am 69.

Chuck Cliburn

ABOUT THE AUTHOR

Chuck Cliburn is a business consultant and lobbyist in Tallahassee, Florida. Prior to founding his consulting firm, New Capitol IT in 2013, he worked for 38 years in the IT industry, including 30 years with tech giants Unisys (previously Sperry Univac and Burroughs) and ACS, a Xerox Company/Xerox. He began his career as a Sales and Marketing trainee with Burroughs in 1974 and completed his corporate career as Senior Vice President and Florida Client Executive for Xerox in 2012.

Before reaching his 30th birthday, Chuck, his wife Madelin, and their son Brian experienced four corporate moves crisscrossing the country from Phoenix to Las Vegas to Miami to Philadelphia to Tallahassee where he held management positions in all these cities. The final corporate move was from Jacksonville to Tallahassee in 1992 where he held senior positions with Unisys and ACS (Xerox) before founding New Capitol IT.

He can be reached at Chuck@NewCapitolLLC.com